CHEMISTRY OF THE CARBONYL GROUP

CHEMISTRY OF THE CARBONYL GROUP

A step-by-step approach to understanding Organic Reaction Mechanisms

Revised Edition

Timothy K. Dickens

Fellow and Director of Studies in Chemistry
Peterhouse, Cambridge

Stuart Warren

Retired Fellow
Churchill College, Cambridge

Registered Office(s)
John Wiley & Sons, Inc., 111 River Street, Hoboken, NJ 07030, USA
John Wiley & Sons Ltd, The Atrium, Southern Gate, Chichester, West Sussex, PO19 8SQ, UK

Editorial Office
9600 Garsington Road, Oxford, OX4 2DQ, UK

For details of our global editorial offices, customer services, and more information about Wiley products visit us at www.wiley.com.

Wiley also publishes its books in a variety of electronic formats and by print-on-demand. Some content that appears in standard print versions of this book may not be available in other formats.

Limit of Liability/Disclaimer of Warranty
In view of ongoing research, equipment modifications, changes in governmental regulations, and the constant flow of information relating to the use of experimental reagents, equipment, and devices, the reader is urged to review and evaluate the information provided in the package insert or instructions for each chemical, piece of equipment, reagent, or device for, among other things, any changes in the instructions or indication of usage and for added warnings and precautions. While the publisher and authors have used their best efforts in preparing this work, they make no representations or warranties with respect to the accuracy or completeness of the contents of this work and specifically disclaim all warranties, including without limitation any implied warranties of merchantability or fitness for a particular purpose. No warranty may be created or extended by sales representatives, written sales materials or promotional statements for this work. The fact that an organization, website, or product is referred to in this work as a citation and/or potential source of further information does not mean that the publisher and authors endorse the information or services the organization, website, or product may provide or recommendations it may make. This work is sold with the understanding that the publisher is not engaged in rendering professional services. The advice and strategies contained herein may not be suitable for your situation. You should consult with a specialist where appropriate. Further, readers should be aware that websites listed in this work may have changed or disappeared between when this work was written and when it is read. Neither the publisher nor authors shall be liable for any loss of profit or any other commercial damages, including but not limited to special, incidental, consequential, or other damages.

Library of Congress Cataloging-in-Publication Data

Names: Dickens, Timothy K., 1957- author. | Warren, Stuart, author.
Title: Chemistry of the carbonyl group : a step by step approach to
    understanding organic reaction mechanisms / Timothy K. Dickens, Stuart
    Warren.
Description: Revised edition | Hoboken, NJ : John Wiley & Sons, 2018. |
    Includes bibliographical references and index. |
Identifiers: LCCN 2017058810 (print) | LCCN 2018007079 (ebook) | ISBN
    9781119459538 (pdf) | ISBN 9781119459552 (epub) | ISBN 9781119459569 (pbk.)
Subjects: LCSH: Carbonyl compounds. | Organic reaction mechanisms.
Classification: LCC QD305.A6 (ebook) | LCC QD305.A6 D53 2018 (print) | DDC
    547/.43–dc23
LC record available at https://lccn.loc.gov/2017058810

Cover Design: Wiley
Cover Image: Courtesy of Timothy K. Dickens

Set in 10/12pt TimesNewRomanMTStd by SPi Global, Chennai, India
Printed and bound in Singapore by Markono Print Media Pte Ltd

10 9 8 7 6 5 4 3 2 1

To Sophie Jackson and
Chris Lester

CONTENTS

5 Building Organic Molecules from Carbonyl Compounds 89

PREFACE

Understanding the movement of electrons as a reaction takes place is perhaps the hardest general concept in Organic Chemistry. This is often referred to as 'pushing curly arrows'. Once this concept has been grasped, it becomes possible to rationalise what is happening in a chemical reaction and predictions can start to be made. In *Chemistry of the Carbonyl Group*, five chemical reactions are explored. These are nucleophilic addition, nucleophilic substitution, nucleophilic substitution with complete removal of carbonyl oxygen, carbanions and enolisation. With these reactions, it is possible to design and build organic molecules from carbonyl compounds. The last section of the book covers this. This understanding of the processes behind reactions by extrapolation can be used to rationalise organic reactions involving heteroatoms such as nitrogen, phosphorus and sulphur. Other types of chemical reactions, such as electrophilic substitution and addition, become easy to comprehend.

It is the authors' firm belief that the most effective way to learn is by practice and interaction. With this in mind, the reader is asked to predict what would happen under a specific set of reaction conditions. The book is divided into frames. These frames pose a question and invite the reader to predict what will happen. Subsequent frames give the solution but then pose more questions to develop a theme further. Therefore, the book should be worked though with pen and paper.

The reactions of the carbonyl group are some of the first reactions that a student studying Chemistry at university will encounter. As such, this book should be tackled just before, or when, a student is starting Organic Chemistry. Indeed, at Peterhouse, first year Natural Science students taking Chemistry are encouraged to work through this book during the Christmas break. Students who do this make substantially faster progress with the Cambridge Organic Chemistry course during the Lent term. The book could also be used by gifted or curious sixth-form students who are keen to broaden their knowledge of Organic Chemistry beyond the A-level syllabus.

This book was first published in 1974. After some discussion, it was decided not to change the text substantially. The motivation was very much to improve the layout of the book; hence *all* the diagrams have been redrawn using ChemDraw and the text formatted using the text mark-up language LATEX. One area that it might have been appropriate to develop is a discussion of the frontier orbitals; this would lead to an understanding of why the "magic angle" of attack in nucleophilic addition[1-3] is 107°. However, this could be seen as an unnecessary distraction, depending on what other Chemistry topics the reader is already familiar with.

Timothy K. Dickens, Cambridge February 2018

[1] I. Fleming. *Molecular Orbitals and Organic Chemical Reactions – Reference Edition* Wiley, (2010). ISBN: 978-0-470-74658-5, section 5.1.3, page 214.

[2] J. Clayden, N. Greeves and S. Warren. *Organic Chemistry.* 2nd Ed. OUP, (2012). ISBN: 978-0-19-927029-3, page 130.

[3] D. Klein. *Organic Chemistry.* 2nd Ed. Wiley, (2015). ISBN: 978-1-118-45228-8, page 937.

ACKNOWLEDGEMENTS

This edition has largely been developed by T.K.D. As the revising author, he is indebted to a number of people, including his son, Alex, who first drew T.K.D.'s attention to it whilst he was assigned this book as a Christmas break exercise when studying Chemistry at New College, Oxford. T.K.D. wishes to express his gratitude to Dr Peter Wothers (Teaching Fellow at the Department of Chemistry at Cambridge) for general advice, to Professor Jonathan Clayden for the discussion on the representation of tetrahedral angles and charges on atoms, to Jenny Cossham, the publisher of this book at Wiley, for her warm encouragement and discussion on layout and presentation of material in this revised edition, to Dr James Keeler (Director of Teaching at the Department of Chemistry) who got the author started with using LATEX and especially to Dr Russell Currie who provided much detailed technical help with the package and for meticulously proof reading drafts of this edition. T.K.D. is indebted to Dr Roger Mallion for the thoroughness with which he checked the manuscript for grammatical and chemical errors. However, the two people whom he would chiefly like to thank are Dr Stuart Warren for all his wonderful help, insight, support and ever helpful encouragement and Catherine, his wife, for her patience every time he disappeared into his study to work on this volume.

SOME HELP THAT YOU MAY NEED

Throughout this book, several references are made to consulting an advisor. An advisor is someone who can guide the reader if a concept is not fully understood or more detail is required. An advisor could be a college tutor or supervisor, lecturer, graduate student or even a student in a later year who has a passion for chemistry. If the book is being tackled by a sixth former, then perhaps his or her chemistry teacher could act in the role of an advisor.

Books that you might find useful:

General Organic Chemistry Textbooks:

J. Clayden, N. Greeves and S. Warren. *Organic Chemistry*. 2nd Ed. OUP, (2012). ISBN: 978-0-19-927029-3.

D. Klein. *Organic Chemistry*. Wiley, (2015). ISBN: 978-1-118-45228-8.

The solutions to the problems posed in these books can be found in:

> J. Clayden and S. Warren. *Solutions Manual to Accompany Organic Chemistry*. OUP, (2013). ISBN: 978-0-19-966334-7.
> D. Klein. *Student Study Guide and Solutions Manual for Organic Chemistry*. 2nd Ed. Wiley, (2015). ISBN: 978-1-118-64795-0.

For those who wish to gain a better grasp of using Molecular Orbitals to describe reactions in Organic Chemistry see:

> I. Fleming. *Molecular Orbitals and Organic Chemical Reactions – Reference Edition*, Wiley (2010). ISBN: 978-0-470-74658-5.

WHAT DO YOU NEED TO KNOW BEFORE YOU START?

The program is about only a small part of organic chemistry, and we have to assume that you know certain facts and appreciate certain concepts. In case you feel uncertain about any of these, here is a list of them with appropriate remedies in each case.[1]

We assume you can:

- Draw and recognise structures of simple organic compounds (an aldehyde, acetone, *n*-butanol, etc.). Any organic textbook will tell you about this.

[1] J. Clayden, N. Greeves and S. Warren. *Organic Chemistry.* 2nd Ed. OUP, (2012). ISBN: 978-0-19-927029-3.

- Write mechanisms using curly arrows (e.g. to described the S_N2 reaction between hydroxide ion and methyl iodide). If you cannot do this, you should read one of the books cited below.[2-3]

- Outline the basic chemistry of alkyl halides including nucleophilic attack at saturated carbon (substitution and elimination reactions S_N1, S_N2, E1 and E2 mechanisms). A brush up with a standard Organic Chemistry textbook might help; see Clayden[4,5] or Klein.[6,7]

- Explain what is meant by the periodic table, pK_a value, anion and cation, electrophile and nucleophile, lone pair of electrons, tetrahedral structure, σ and π-bonds, intra-molecular and reversible reactions. Again, any organic text will help.

This list applies to the whole program; however, at the beginning of each section, you will find a list of *Concepts Assumed*. This will be a more detailed analysis of the concepts we expect you to have grasped before you start that section. You could find help in textbooks or from your adviser if you are unsure of any of these concepts.

There are also lists of *Concepts Introduced* and *Concepts Reinforced* for each section. You may find these useful as a checklist to make sure, after you have finished the section, that you really have met and understood these concepts. In many cases, a concept introduced in one section will be assumed for the next.

---

[2] J. Clayden. N. Greeves and S. Warren. *Organic Chemistry.* 2nd Ed. OUP, (2012). ISBN: 978-0-19-927029-3. chapter 5, page 107.

[3] D. Klein, *Organic Chemistry.* 2nd Ed. Wiley, (2015). ISBN: 978-1-118-45228-8, chapter 3, page 103.

[4] J. Clayden. N. Greeves and S. Warren. *Organic Chemistry* 2nd Ed. OUP, (2012). ISBN: 978-0-19-927029-3, Chapters 15 and 17.

[5] J. Clayden. N. Greeves and S. Warren. *Organic Chemistry* 2nd Ed. OUP, (2012). ISBN: 978-0-19-927029-3, chapter 17.

[6] D. Klein, *Organic Chemistry.* 2nd Ed. Wiley, (2015). ISBN: 978-1-118-45228-8, Chapters 7 and 8.

[7] D. Klein, *Organic Chemistry.* 2nd Ed. Wiley, (2015). ISBN: 978-1-118-45228-8, chapter 8.

INTRODUCTION

HOW TO USE THE PROGRAM

Remember at all times that the point of program learning is that you learn at your *own pace* and that you yourself check on your *own progress*. We shall give you information and ideas in chunks called *frames*, each numbered and separated by a blue line. Normally, each *frame* includes a question, which is sometimes followed by a comment or a clue, and always by the answer. A program is an active form of learning: if it is to be of any use to you, you must play your part by actually *writing down the answers to the questions* as you go along and checking up on points you are not sure about.

When you are ready to start, cover the first page with a card and pull it down to expose the first *frame*. Read and act on that *frame* and then expose *frame* **2**, and so on. Remember to write down the answers to the questions – they are for you to check on your own progress and it is often only when you commit your answer to paper that you find out whether you really understand what you are doing.

Throughout this book, the following colour conventions have been followed in diagrams:

Red   arrows imply the movement of electrons. **Red** has also been used to highlight formal negative charge and lone pairs of electrons.

Green   has been used to denote formal positive charge.

Blue   has been used to show free rotation around a single bond where it is pertinent to the discussion.

Where molecules are explicitly referenced in the text they are labelled in **Blue**. References to frames are also labelled in **Blue**. Positions of carbon atoms relative to each other, where explicitly discussed, are labelled in **Red**.

We hope you find the program enjoyable and helpful.

CHAPTER 1

NUCLEOPHILIC ADDITION TO THE CARBONYL GROUP

Contents

Chemistry of the Carbonyl Group:
A step-by-step approach to understanding Organic Reaction Mechanisms,
Revised Edition. Timothy K. Dickens and Stuart Warren.
© 2018 John Wiley & Sons Ltd. Published 2018 by John Wiley & Sons Ltd.

Concepts assumed:

- σ- and π-bonds.
- Polarisable bonds.
- Electrophile and nucleophile.
- Conjugation with π-bonds and lone pairs.
- Inductive effects.
- pK_a values.
- Periodic table.
- Transition states.

Concepts introduced:

- Acid catalysis.
- Instability of $R_2C(OR)_2$ in acid solution. Driving equilibria in a chosen direction by the use of acid, solvent, etc.
- Stability of different carbonyl compounds. Stability and reactivity as two sides of the same coin.
- Effect of substituents on equilibria. Relationship between basicity and nucleophilicity.
- Organo-magnesium compounds as nucleophiles.
- Use of Grignard reagents in syntheses.
- Ease of dehydration of tertiary alcohols in acid solution.
- Sources of nucleophilic H^-.
- Use of Al and B compounds and as anion transferring reagents.
- Drawing transition states.
- Stability of the six-membered ring.
- Use of protecting groups.
- Use of reaction mechanisms in syntheses.

Have you read the introductions explaining what help you need, what you need to know and how to use the program? It's a good idea to do this before you start.

---

1. The carbonyl group **(1a)** has an easily polarisable π-bond with an electrophilic carbon atom at one end easily attacked by nucleophiles:

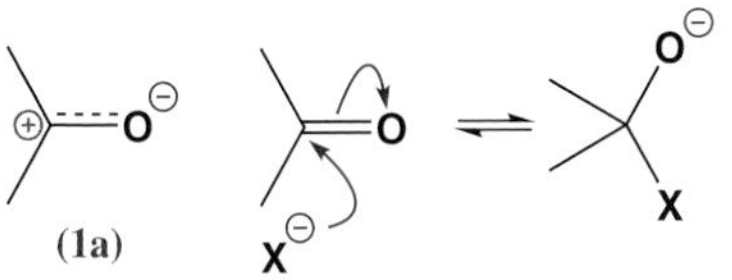

(1a)

Write down the reaction (with curly arrows) between acetone and hydroxide ion.

---

2. Have you actually written down the formulae of the reagents and drawn the arrows? The program won't be of much help to you unless you do.

3.

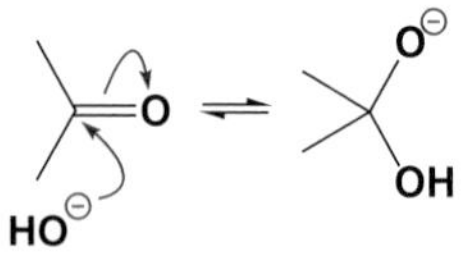

A nucleophile such as water uses its lone pair electrons (:) to attack and forms a neutral addition compound by proton transfer:

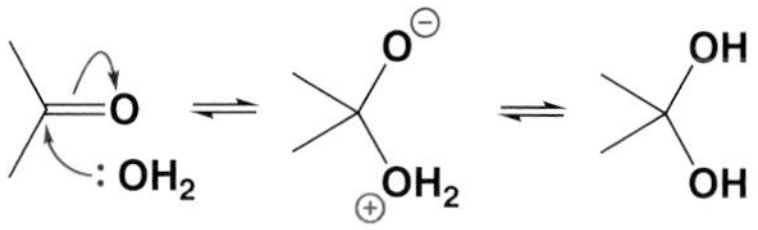

Note that only one proton is needed.

Write down the reaction between the carbonyl compound acetaldehyde and the proton-bearing nucleophile ethanol.

___

4.   If you find this difficult, use the lone pair electrons on the ethanol oxygen atom to attack the carbonyl group of acetone.

___

5. Answer to *frame* **3**:

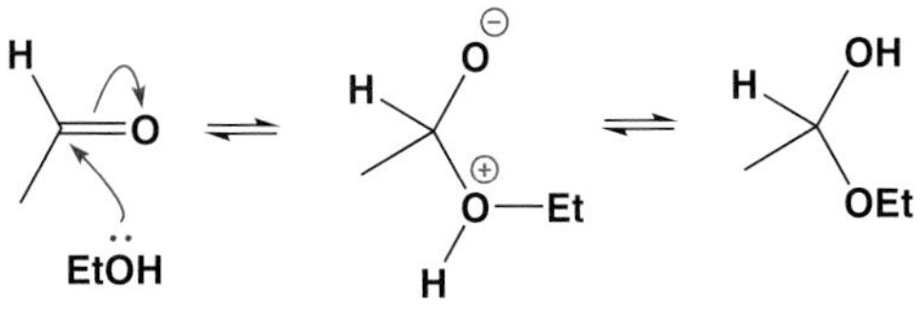

Another approach is to add the proton first, in acid solution, and to add the nucleophile afterwards:

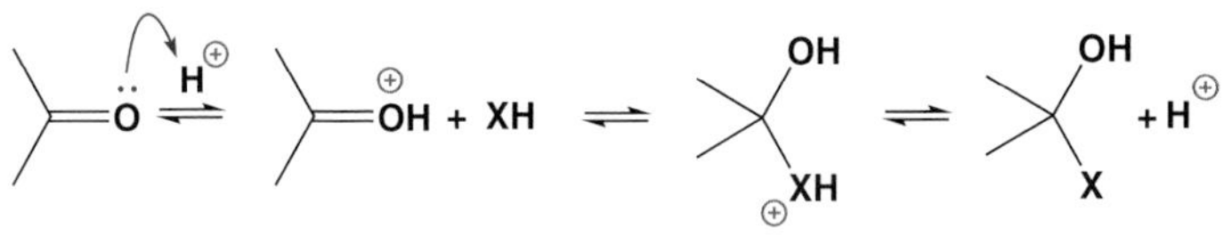

In this case, the proton is regenerated and this is an example of acid catalysis. Show how water can be added to acetone with acid catalysis.

_____

6.   If you are having difficulty with this, look back at the last reaction in *frame* **5**. Carry out these same steps using acetone as the carbonyl compound and water as HX.

_____

7.

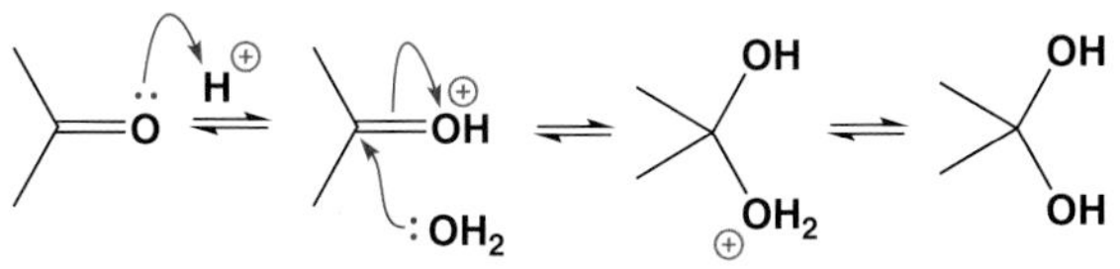

Notice that $Me_2C=O^+H$ is much more reactive than acetone, but is still attacked at *carbon* although the positive charge is in fact on the oxygen atom. What would happen to $R_2C=O^+H$ with water, ethanol, $PhCH_2SH$ and cyanide ion?

_____

8. The compounds:

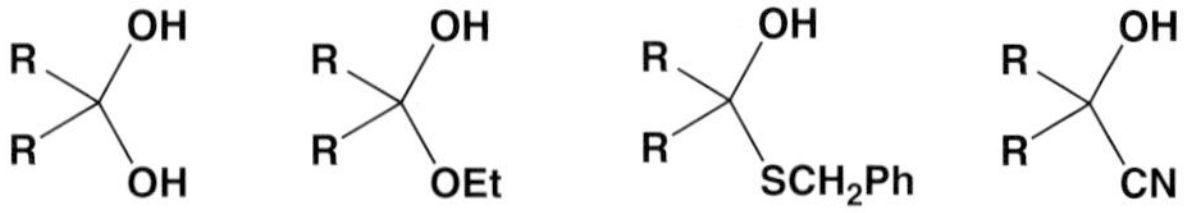

(8b)

would be formed by a mechanism exactly like the one in *frame* **7**. When you combined $R_2C=O^+H$ with ethanol you formed an adduct **(8b)**, which is the product of ethanol addition to a ketone. The steps you drew are therefore part of the acid catalysed addition of ethanol to a ketone. Draw out the whole of this reaction.

9.

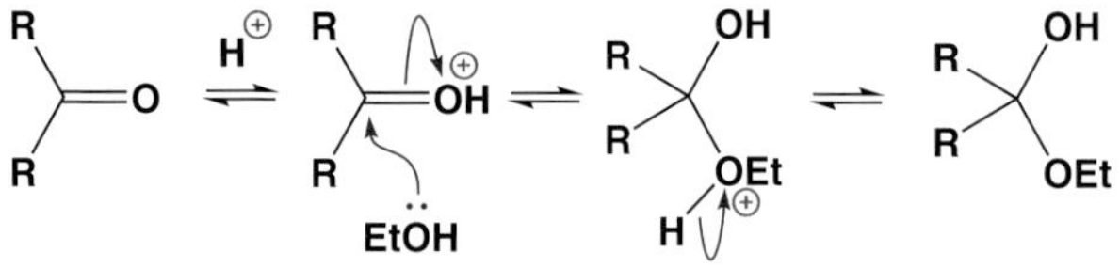

10.   Look at the reactions in *frames* **7** and **9** again. Note that all the steps are reversible and that therefore $R_2C=O^+H$ may be formed from $R_2C=O$, $R_2C(OH)_2$ or $R_2C(OH)OR$:

(10c)

It is in fact a general rule that compounds of the type $R_2C(OR)_2$, having two oxygen atoms singly bonded to the same carbon atom, are unstable in acid solution. A reason for this is that both oxygens have lone pairs of electrons, and so, when one pair is protonated, the lone pair on the other can form a C=O double bond and expel the protonated atom. Draw this in detail.

11.

Look at the reactions in *frame* **10** again. In the reverse reaction, we protonated and removed the EtO⁻ group from **10c**. What happens if we protonate and remove the HO⁻ group?

12.

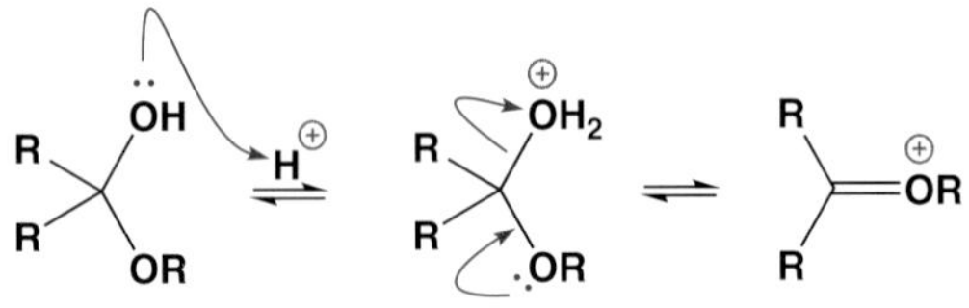

This new cation, $R_2C=O^+Et$, is just as reactive as $R_2C=O^+H$ and can add nucleophiles in the same way. What happens if we add EtOH to it?

13.

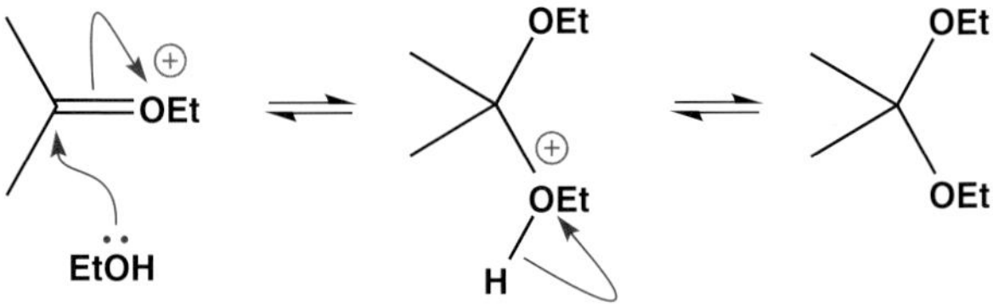

This reaction sequence, added to the ones in *frames* **9** and **12**, gives us the addition of two molecules of ethanol to a ketone to give $R_2C(OEt)_2$. Draw this sequence out in full without referring back.

---

14. If you have difficulty doing this, look at *frames* **9**, **12** and **13** without writing anything down and then try again.

15.

---

16.   Does your reaction sequence exactly follow that in *frame* 15? If not, try to assess if the differences are trivial. If you are still in doubt, consult your adviser. It is important that you understand this reaction well.

This is a good place to stop if you want a break.

---

17.   Since this whole sequence is reversible, it will go forwards in ethanol and backwards in water. What do you think would happen if acetaldhyde and *n*-butanol were dissolved together in a 1 : 3 molar ratio, and the solution refluxed for 12 hours with a catalytic quantity of toluene sulphonic acid, and the product dried and distilled?

---

18.   This is a literature preparation of $CH_3CH(OBu^n)_2$. Acetaldehyde is the carbonyl component, butanol is the nucleophile and toluene sulphonic acid is the catalyst. How would you hydrolyse $PhCH(OEt)_2$, and what would you get?

---

19. Reflux $PhCH(OEt)_2$ in water with a catalytic quantity of an acid. The products would be PhCHO and EtOH. How would you make $EtCH(OMe)_2$?

---

20. Treat EtCHO and MeOH as in *frame* **17**. A glance at the reactions in *frame* **15** should convince you that the *mono* adducts of carbonyl compounds and nucleophiles with lone pairs are unstable. An example is $R_2C(OH)OEt$, which is unstable even under the conditions of its formation. What happens to it?

---

21. In ethanol it gives the acetal $R_2C(OEt)_2$, in water the carbonyl compound $R_2C=O$.

---

22. If we look instead at nucleophiles *without* lone pairs, we should find some stable *mono* adducts. Which of the adducts in *frame* **8** should be stable?

---

23. The cyanide substituent has no lone pair and so its adduct, $R_2C(OH)(CN)$, should be stable. These compounds, cyanohydrins, are made by adding excess NaCN and one equivalent of acid to the carbonyl compound. The reaction is an equilibrium. What is the role of the acid?

---

24. To drive over the equilibrium by protonating the intermediate:

---

25. Since this reaction is an equilibrium, the amount of cyanohydrin formed from any given carbonyl compound will depend on the relative stabilities of the carbonyl compound itself and the product. There can be many substituents 'X' on a carbonyl compound RCOX, such as Cl, Me, NH$_2$, Ph, OEt, H. Some have inductive effects, some are conjugated with the carbonyl group. Some stabilise RCOX making it less reactive. Others activate it towards nucleophilic attack. Arrange the compounds RCOX, where 'X' can be the substituents listed above, into an order of reactivity towards a nucleophile.

---

26. Before you look at the answer in the next frame, just check that you have considered each of these factors: some substituents stabilise RCOX by π-conjugation:

some by lone pair conjugation:

some destabilise RCOX by inductive electron withdrawal:

---

27. Taking RCHO as standard, we can say that Cl destabilises by inductive withdrawal, Me stabilises more by σ-delocalisation and NH₂ and OEt by lone pair donation. (NH₂ is more effective at this: compare ammonia and water as bases). Our order is

$$\text{Cl} \ldots \text{H} \ldots \text{Me} \ldots \text{Ph} \ldots \text{OEt} \ldots \text{NH}_2$$
most reactive   *most stable*

These same factors could affect the product as well. Arrange the same substituents in an order for product stability.

---

28. Since the carbonyl group has gone in $R_2C(OH)CN$, we would expect very little effect from any of these substituents. There cannot be any conjugation, and inductive effects on the distant O atom will be small. What effect would an inductively withdrawing substituent have on the equilibrium for cyanohydrin formation?

---

29. It will destabilise the carbonyl compound a lot and have very little effect on the product; it will therefore push the equilibrium over to the cyanohydrin side. Consider cyanohydrin formation from:

(a)   MeCHO   (c) $Me_2C=O$ (e) $Ph_2 C=O$
(b)   PhCHO   (d) $Ph(CO)=Me$ (f) $Me(CO)OEt$

Two of these give no cyanohydrin at all. Which two? One gives 100% cyanohydrin — which? When equilibrium is established, (c) forms 10 000 times as much cyanohydrin as (d). Comment.

---

30. (e) and (f) give no cyanohydrin at all: (a) gives all cyanohydrin. In (d) there is a strong π-conjugation from the benzene ring absent in (c) that stabilises the carbonyl compound. These results fit in with what we have said; if you don't see, consult your adviser.

---

31. Another carbon anion you may have met is the acetylide ion, formed by the action of strong base on acetylene:

$$H \equiv\!\!\!-\!\!\!-H \quad \overset{\ominus}{NH_2} \longrightarrow H \equiv\!\!\!-\!\!\!\ominus + NH_3$$

It adds readily to the carbonyl group. Draw out the reaction.

32.

33. The pK_a of HCN is 9.15, that of water is 15.7 and that of acetylene is about 25. Which anion, CN^-, HO^- or acetylide ion, would add most rapidly to acetone?

34. Acetylide fastest, then hydroxide, then cyanide slowest. Remember that if HX is a weak acid, X^- is a strong base.

35. Another type of carbon nucleophile is the Grignard reagent RMgBr made by direct metalation of the organic halide with magnesium metal. These compounds are nucleophilic through carbon because the electrons in the C—Mg bond polarise towards carbon. Draw the attack of MeMgBr on CH_2=O.

36.

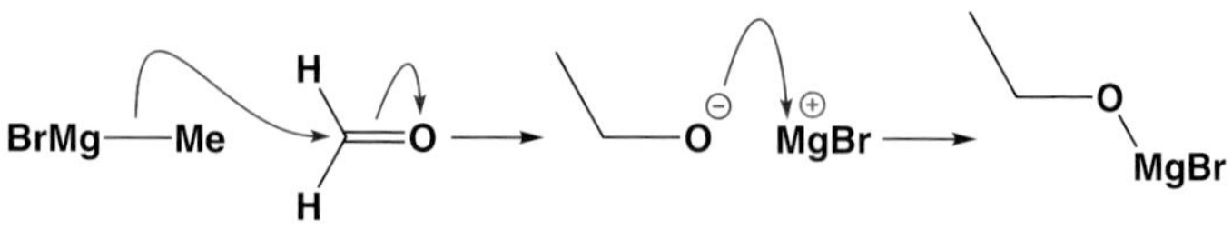

The intermediate O—Mg compound hydrolyses in acid solution by attack of water on the magnesium atom. Suggest how this might occur.

---

37.   A possible mechanism is:

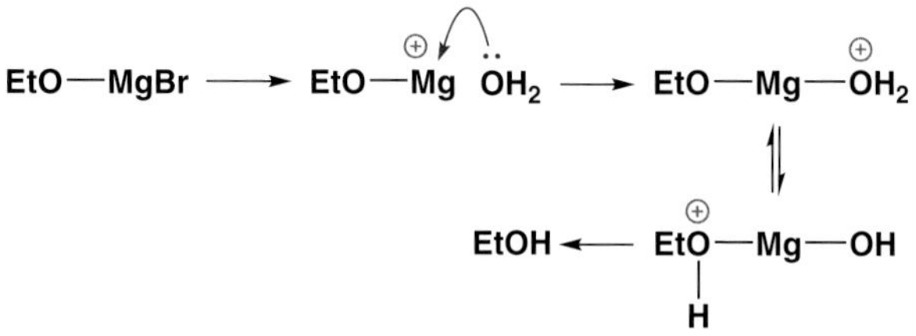

In any event, the product is a primary alcohol and is a general route from RBr to RCH_2OH. What would be the reaction between $PhCH_2MgBr$ and CH_3CHO?

38.

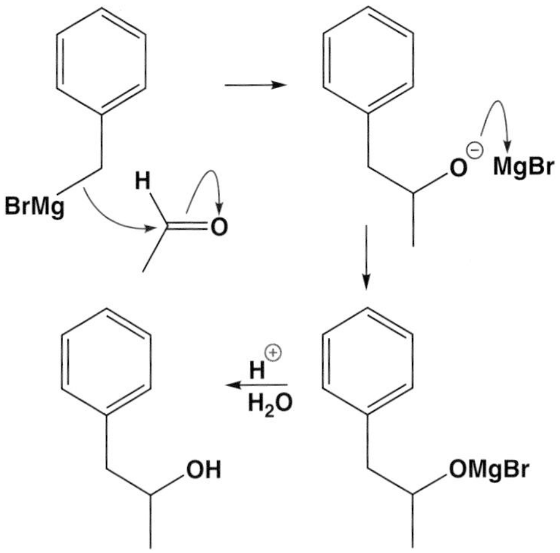

We can therefore make secondary alcohols this way. Ketones also react with Grignard reagents. Draw the reaction between $PhCH_2MgBr$ and $Ph_2C=O$.

39.

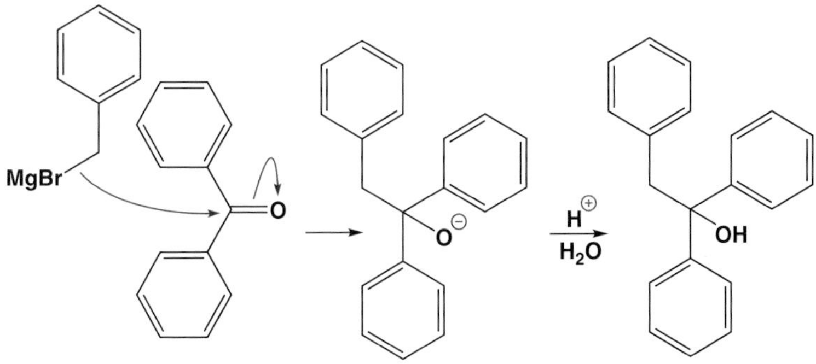

The work-up is done in acid solution, and tertiary alcohols react easily with acid. What further reaction might happen here?

---

40. Protonation and loss of water lead to a stable tertiary carbonium ion:

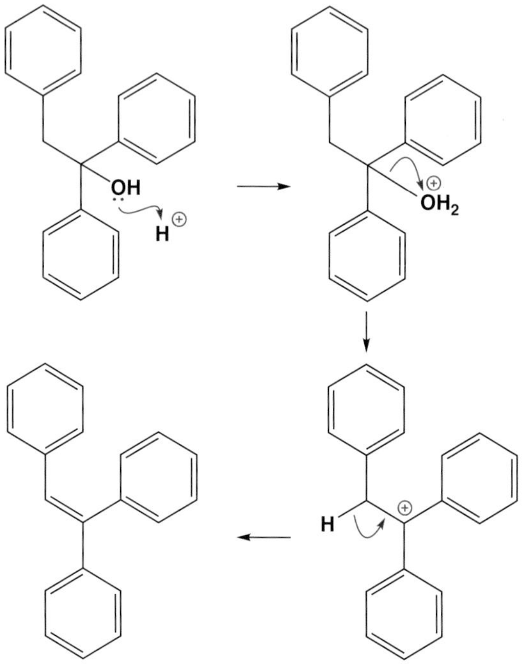

So much for carbon nucleophiles.

This is a good place to rest.

---

41. The simplest nucleophile of all is the hydride ion, H^-, but, as you may know, this ion is very basic and will not add to the carbonyl group. Sources of H^- for addition to the $C=O$ bond are $NaBH_4$ and $LiAlH_4$ containing the tetrahedral anions BH_4^- and AlH_4^-. Draw out the structure of these ions.

42.

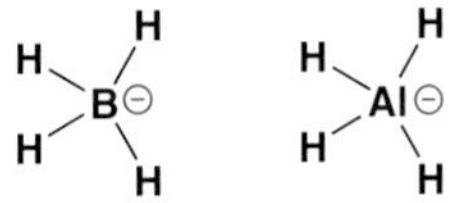

Show how AlH_4^{-} may transfer H^{-} as a nucleophile to a ketone, $R_2C{=}O$.

43.

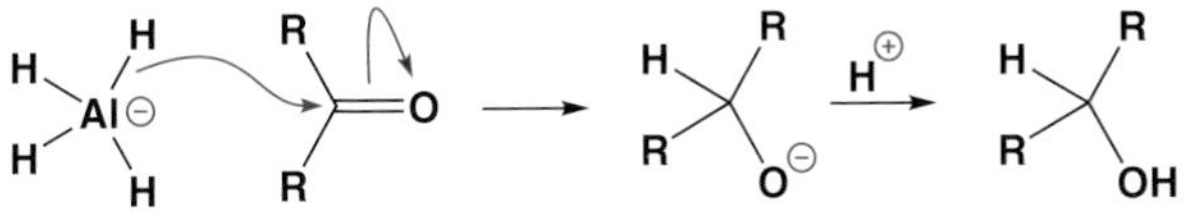

Lithium aluminium hydride, $LiAlH_4$, is dangerous when damp. Can you suggest why?

44.   It gives off hydrogen in large volumes ($H^{-} + H_2O \longrightarrow H_2 + OH^{-}$) and evolves heat at the same time. The result is usually an impressive fire. Sodium borohydride, $NaBH_4$, is less reactive and can be used in alkaline aqueous solution. What would be formed from $PhCOCH_3$ and $NaBH_4$?

45.

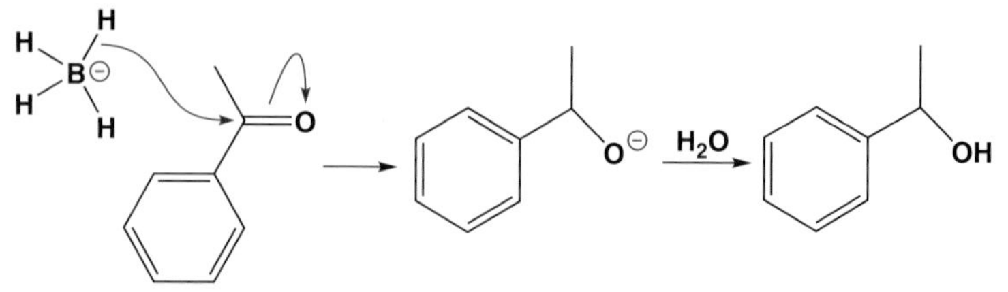

These reagents demonstrate two important properties of boron and aluminium compounds. If you are uncertain of the Periodic Table, just check to see where these two elements come. Neutral tervalent B and Al compounds are electron deficient: only six valency electrons: no lone pair.

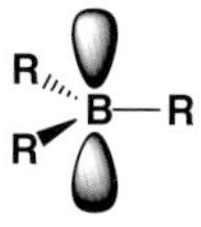

They readily accept nucleophiles to form stable tetravalent anions. Draw the reaction between BF_3 and a fluoride ion, F^-

_____

46.

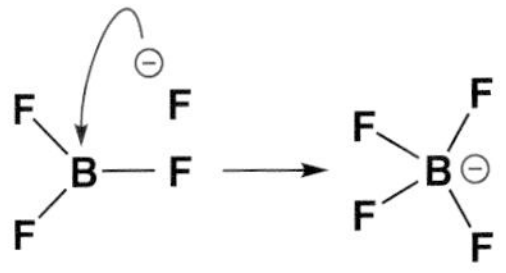

Also, anions can be transferred from these tetrahedral anions to other molecules:

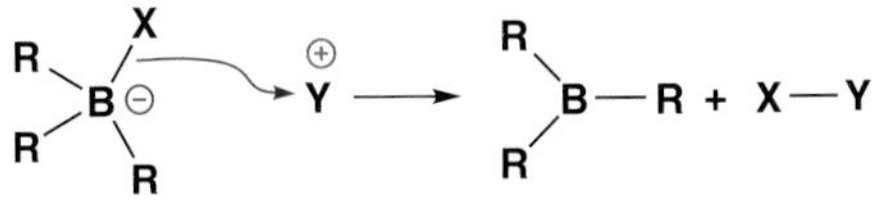

These two properties may be summarised by saying that tervalent boron and aluminium compounds will accept anions from one molecule and transfer them to another.

---

47. This is used in the selective reduction technique known as the Meerwein–Ponndorf reduction. The tervalent compound is aluminium *iso*-propoxide, $(i\text{-}PrO)_3Al$. When a compound such as a ketone is added to this reagent, it combines with it to form a tetrahedral anion. Draw this.

---

48. If you are in difficulty, remember that the Al atom is electrophilic and therefore combines with nucleophiles, and think which end of the ketone molecule is nucleophilic.

49.

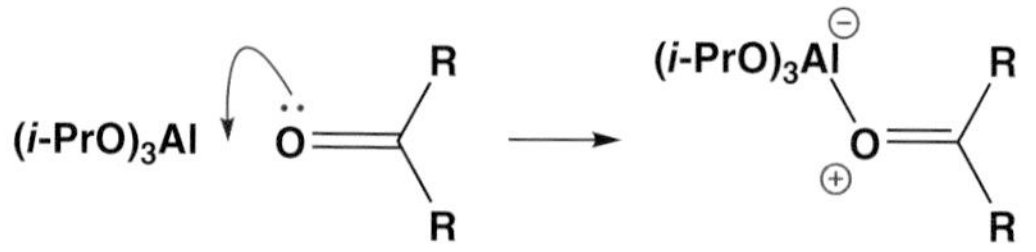

If we draw this same intermediate with one of the *iso*-propyl groups drawn out in full, we can see that the tertiary hydrogen atom in the *iso*-propyl group (*) is geometrically placed so that it can be transferred to the ketone. Put arrows on the formula to show how this happens.

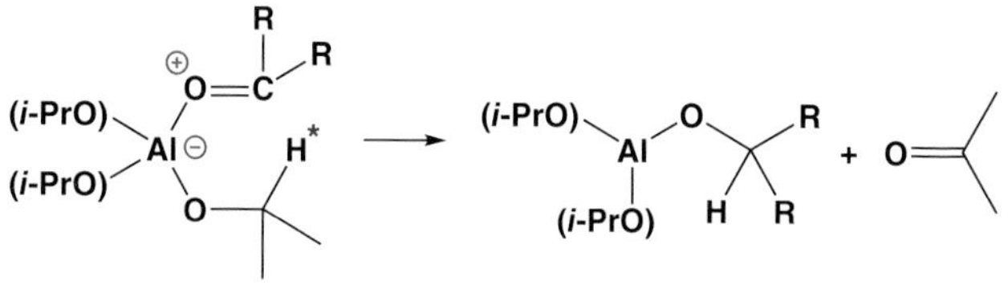

50.

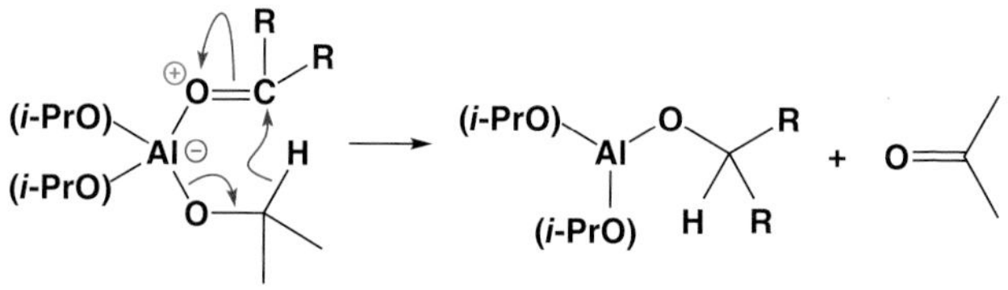

51. The products so far are acetone and a new aluminium compound. The reaction is done in *iso*-propanol solution so that another *i*-PrO group displaces the newly formed alcohol from the Al atom:

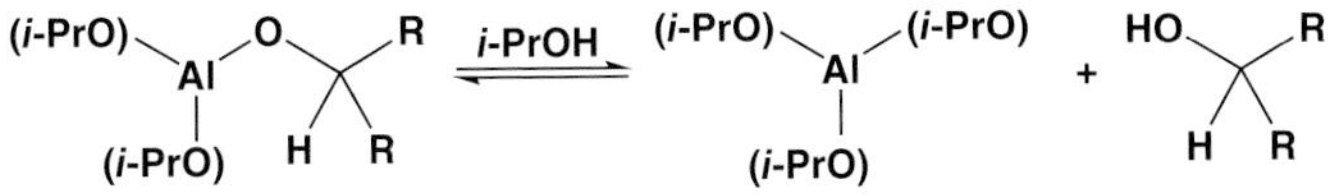

The reaction is done at a high enough temperature for the acetone to distil off and the equilibrium is kept over to the right.

---

52. In the hydrogen transfer reaction (*frame* **50**), how is the hydrogen actually transferred – as an atom, a proton or a hydride ion?

---

53. As a hydride ion – that is, with the pair of electrons from the C—H bond. This type of reaction is called a hydride transfer. You may have noticed that the transfer was done *intra*-molecularly within a six-membered ring, and therefore it has a six-membered transition state. Draw it.

---

54. If you do know how to draw transition states, skip to *frame* **59**. If you don't know, read on. The transition state for a reaction is the state of highest energy along the reaction path. To draw it, one must first know the reaction mechanism in detail. If there are more than one step, then there will be a transition state for each step. We shall go through the process for a one-step reaction, BH_4^- and acetone. Draw the mechanism for this reaction.

55.

$$BH_3 - H \quad \diagup = O \longrightarrow H_3B + \diagdown{O}^\ominus$$

Now draw the formulae again, but put in only those bonds that remain unaffected by the reaction. Don't draw charges or arrows.

56.

$$H_3B \quad H \diagdown - O$$

Now 'dot' in all bonds that are formed or broken during the reaction and mark all appearing or disappearing charges in brackets to show that they are partial charges. (Don't use δ+ or δ− as these mean very small charges and here the total charge must add up to unity.)

57.

$$\overset{(-)}{H_3B} ---- H ----- \diagdown = \overset{(-)}{O}$$

Now draw the transition state for this reaction:

$$\underset{R}{\diagup}\overset{O}{\diagdown}_{O} H \quad = N = N \longrightarrow \underset{R}{\diagup}\overset{O^\ominus}{\diagdown}_{O} + \quad -N \equiv N$$

58.

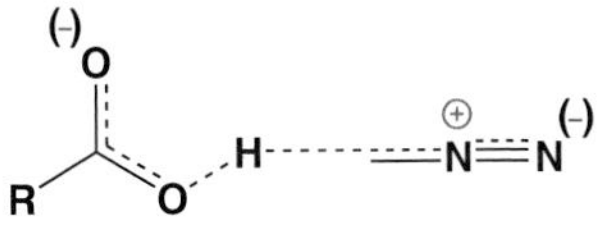

If this wasn't clear, consult your adviser. If it was clear, you ought to be able to do the original problem so go back to *frame* **53**.

59.

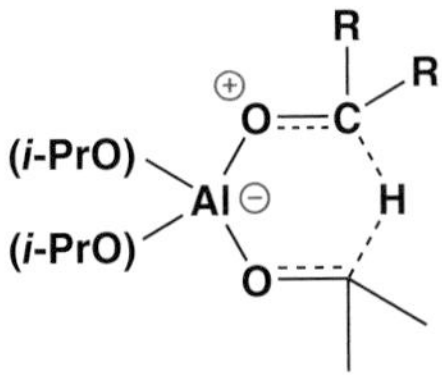

The reason that this reaction goes so well is that six-membered rings are very stable, and so an intramolecular reaction going through a six-membered transition state will be the most favourable.

---

60.   The reverse reaction is known as the Oppenauer oxidation. Here, aluminium-tri-*tert*-butoxide and an involatile ketone such as cyclohexanone are used to oxidise any secondary alcohol to the corresponding ketone.

Show how a hydride transfer in this intermediate leads to the oxidation of the alcohol.

61.

We can reduce a ketone to an alcohol or oxidise an alcohol to a ketone by using $(i\text{-PrO})_3\text{Al}$ and acetone in the one case and $(\text{Bu}^t\text{O})_3\text{Al}$ and cyclohexanone in the other. By their mechanism you can see that these reactions will have no effect on other functional groups such as C=C double bonds.

---

62. This is nearly the end of the first part of the program, so here are some general problems. You will remember that acetal formation (*frames* 7–17) is a reversible reaction. It turns out that the equilibrium constant for acetal formation from a ketone is unfavourable:

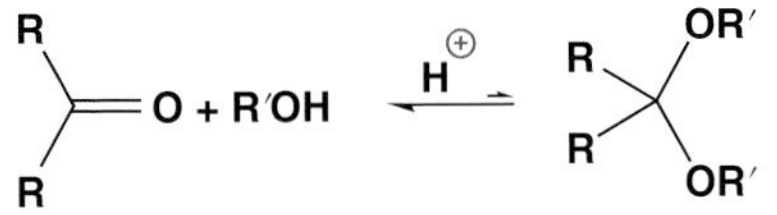

and poor yields are obtained. However, cyclic acetals can be made from ethylene glycol. Draw out the mechanism for this reaction:

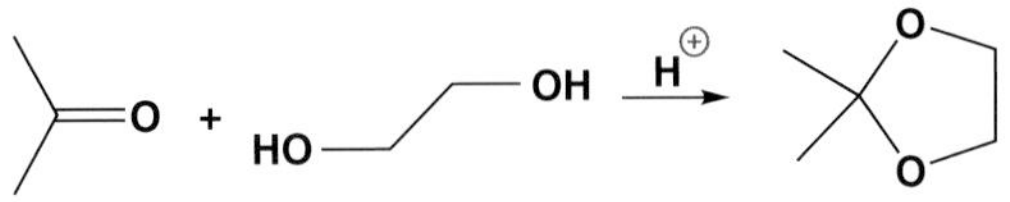

63. If you need some help, begin by adding one OH group of $HOCH_2CH_2OH$ to the protonated ketone, just as you did with ethanol in *frame* **5**.

64.

Predict what happens in this reaction sequence:

65.   In the first step, the acetal from the aldehyde group but not from the ketone group (*frame* **62**) is formed:

Only one carbonyl group is now available for Meerwein–Ponndorf reduction (*frames* **47–61**).

Finally, the acetal is hydrolysed by standard means. The result of all this is that we have reduced a ketone in the presence of an aldehyde:

This is the end of the first part of the program.

CHAPTER 2

NUCLEOPHILIC SUBSTITUTION

Contents

Chemistry of the Carbonyl Group: **29**
A step-by-step approach to understanding Organic Reaction Mechanisms,
Revised Edition. Timothy K. Dickens and Stuart Warren.
© 2018 John Wiley & Sons Ltd. Published 2018 by John Wiley & Sons Ltd.

Concepts assumed:

- Electrostatic repulsion.
- Catalysis.
- Heavy isotopes.
- Acidity and basicity of Cl^-, EtO^-, NH_3, etc.

Concepts introduced:

- Relationship between nucleophilicity, leaving group ability and basicity.
- Substitution as an extension of addition.
- Position and use of electrophilic attack on carboxylic acid derivatives.
- Similarity of reaction between cyclic and acyclic compounds.
- Why carboxylic acids are unreactive towards nucleophilic substitution.
- Use of ^{18}O in establishing reaction mechanism.

Concepts reinforced:

- Use of acid and base catalysis and choice of solvent in driving an equilibrium in the chosen direction.
- Advantages of the six-membered cyclic transition state.
- Use of Grignard reagents in synthesis.
- Ease of dehydration of tertiary alcohols.
- Use of reaction mechanisms in designing syntheses.

66. In the first part of the program, we considered only aldehydes and ketones. We're now going to look at the full range of structures including carboxylic acids, R(CO)OH, acid chlorides R(CO)Cl and esters R(CO)OR.

Using your knowledge of nucleophilic addition, what would be the first reaction between hydroxide ion and benzoyl chloride, Ph(CO)Cl?

67.

Instead of picking up a proton from the solvent, this intermediate has a better reaction: the negative charge returns to restore the C=O double bond expelling Cl⁻. Draw this.

68.

Why did it expel Cl⁻ and not OH⁻?

69. Because Cl⁻ is a better leaving group than OH⁻. We know this because HCl is a stronger acid than HOH and is therefore readier to ionise. Cl⁻ must therefore be more stable than OH⁻. Draw out the whole of the reaction.

70.

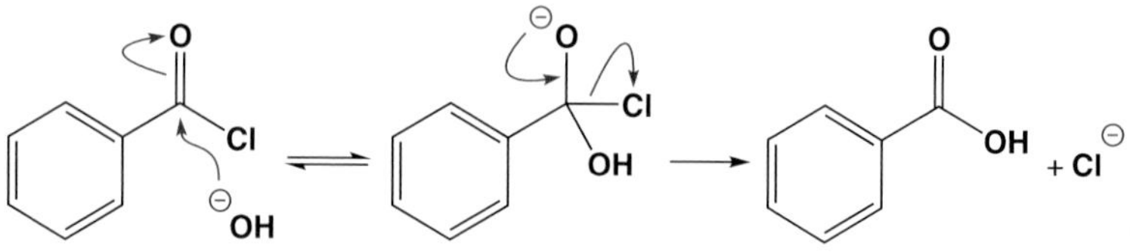

This is then a substitution reaction, OH⁻ in the product taking the place of Cl⁻ in the reactant.

---

71.   What happens if aniline, $PhNH_2$, and benzoyl chloride react together?

---

72.   Aniline is the nucleophile:

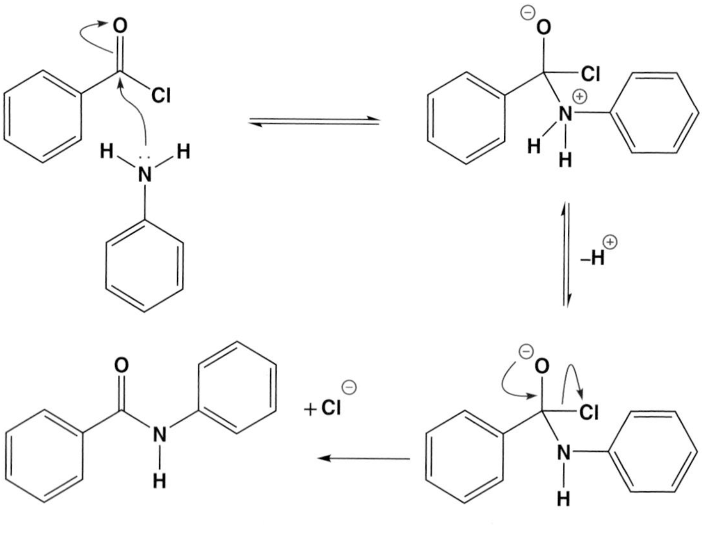

Cl⁻ is preferred to PhNH⁻ as leaving group. Which is the stronger nucleophile towards a carbonyl group, Cl⁻ or OH⁻?

---

73. OH⁻ because it is more basic. It is in fact a general principle that, as far as the carbonyl group is concerned, molecules are better nucleophiles if they are more ____ _____ and worse leaving groups if they are more ____ _____.

---

74. *Basic* fills both gaps. To put it another way, a more/less basic nucleophile will displace a more/less basic leaving group from a carbonyl compound. (Choose 'more' or 'less' in each case.)

---

75. More. Less. Which of these reactions would you expect to work well?

(a) $Cl^- + CH_3(CO)NH_2 \longrightarrow CH_3(CO)Cl + NH_2^-$
(b) $NH_3 + CH_3(CO)OEt \longrightarrow CH_3(CO)NH_2 + EtOH$
(c) $CH_3(CO)O^- + CH_3(CO)Cl \longrightarrow CH_3(CO)O(CO)CH_3 + Cl^-$

---

76. (b) and (c) will work, (a) won't. If you don't see this, consult your adviser. What do you think will happen here?

$$R(CO)OEt + LiAlH_4 \longrightarrow$$

---

77. If you have problems, read *frames* **42** and **43** to remind yourself that the AlH_4^- ion delivers H^- as a nucleophile to the carbonyl group.

78.

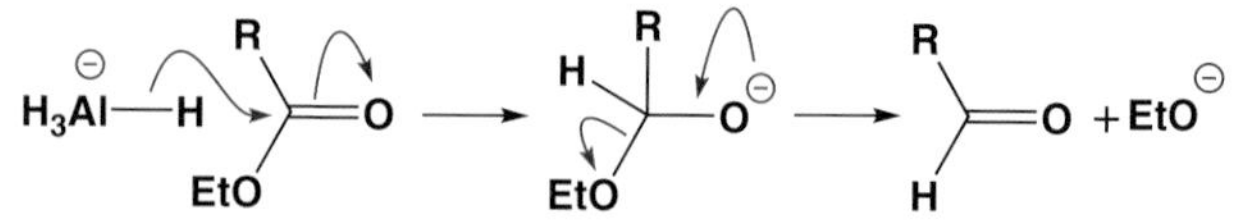

EtO^- is less basic than H^- and so is displaced. What will happen to the product under the reaction conditions?

---

79.   The aldehyde is reduced to the alcohol:

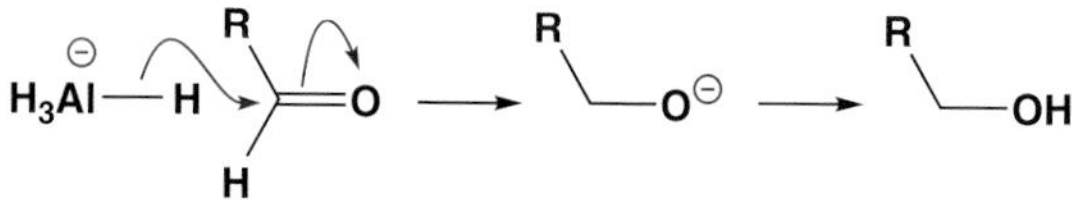

Since the aldehyde is normally more reactive than the ester, it is virtually impossible to stop at the first stage and the compound is normally treated with an excess of $LiAlH_4$ to give the alcohol.

---

80.   A very important series of nucleophiles are the Grignard reagents, $RMgBr$. What would happen with $R(CO)OEt$ and $R'MgBr$? The reaction begins like this:

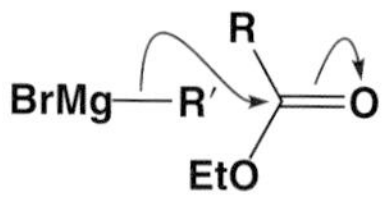

---

81.   Have you considered the possibility of further reaction? *Frames* **35–38** may help.

---

82. As before, the first reaction converts these carbonyl compounds into reactive products, this time ketones. An excess of Grignard reagent therefore gives good yields of tertiary alcohols.

The double-headed arrow on the ester molecule is a useful shorthand to indicate the two stages of the substitution reaction. We shall use it from now on, and if you are in doubt about its meaning you should consult your adviser. The final product from this reaction was a tertiary alcohol. How might this be made into an olefin?

---

83. Tertiary alcohols give carbonium ions with great ease in acid solution (see *frames* **39–40**), and the carbonium ion can either pick up a nucleophile or lose a proton to give an olefin. To get a good yield of olefin you therefore want an acid with a weakly nucleophilic anion, such as H_2SO_4 or $KHSO_4$. In fact this reaction often happens during the workup of the Grignard reaction.

---

84. What product would be formed in this reaction?

85.   Here we used a cyclic substituent.

Now try this reaction with a cyclic ester or lactone in it. The reaction is just like that for an ordinary ester.

86.

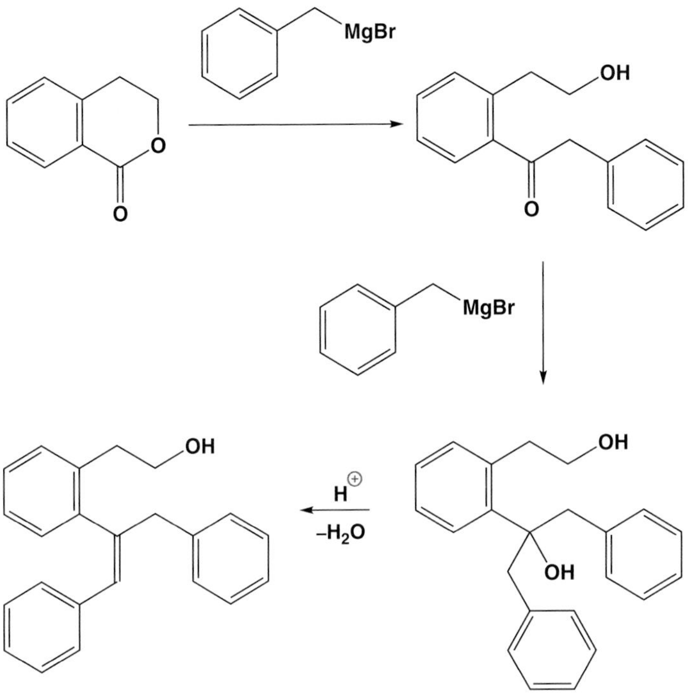

87. A Grignard reaction is probably an example of electrophilic catalysis involving two molecules of the Grignard reagent:

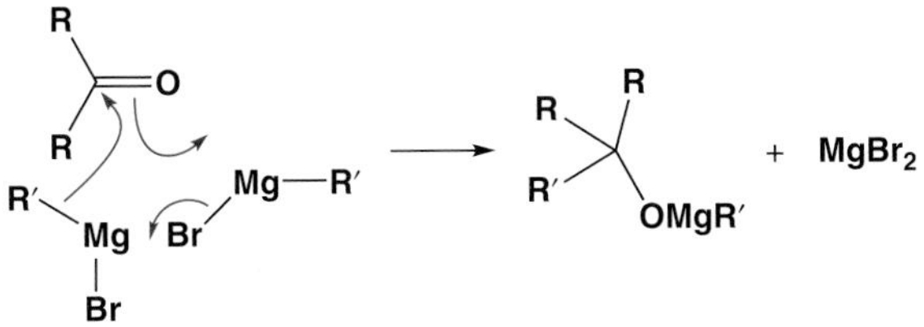

This intramolecular mechanism may remind you of the Meerwein–Ponndorf reduction (*frames* **49–59**). Both have a six-membered cyclic transition state. Draw it for this reaction.

88.

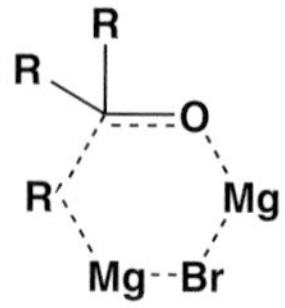

This is a good place to rest if you want to.

---

89.   In view of our conclusion in *frame* **74**, it may surprise you to recall that esters are hydrolysed in alkaline solution:

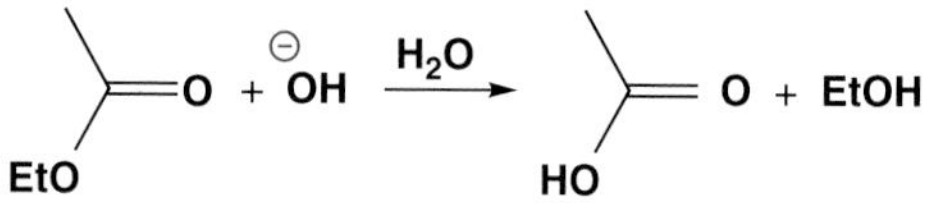

Draw a mechanism for this.

90.

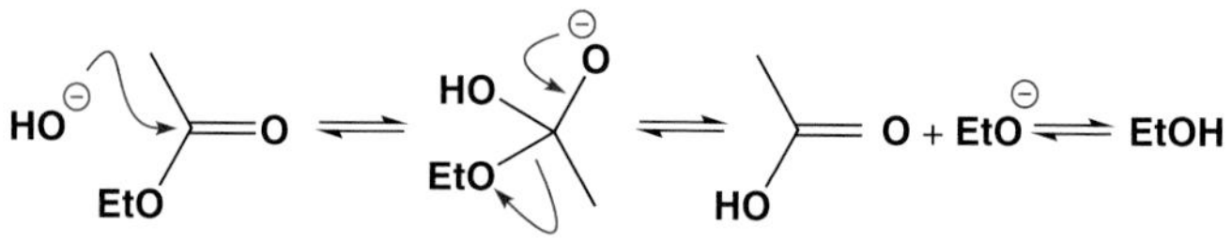

Here, OH⁻ displaces the slightly more basic EtO⁻. Can you think why this happens so easily?

---

91.   A clue: did you notice that the steps are all equilibria? Think how an equilibrium can be driven over, and perhaps compare this reaction with the hydrolysis of acetals (*frame* **17**).

---

92. The reaction is done in water with an excess of OH⁻ to drive the reaction forward by the mass-action effect. There is another reason too. Think which species will actually be formed in aqueous alkali.

---

93. $CH_3(CO)OH$ will form $CH_3(CO)O^-$ and electrostatic repulsion will prevent attack of OH⁻ on this so that it is effectively removed from the equilibrating system.

---

94. Amides can also be hydrolysed in alkali, but let's look at their hydrolysis in acid solution:

$$\underset{H_2N}{\diagdown}{=}O + H_2O \xrightarrow{\overset{\oplus}{H}} \underset{HO}{\diagdown}{=}O + NH_3$$

How could acid catalyse the first step of the reaction, the formation of the tetrahedral intermediate?

---

95. By protonation. Note that we use the lone pair of electrons on nitrogen, but protonate at the carbonyl group to get the most delocalised cation:

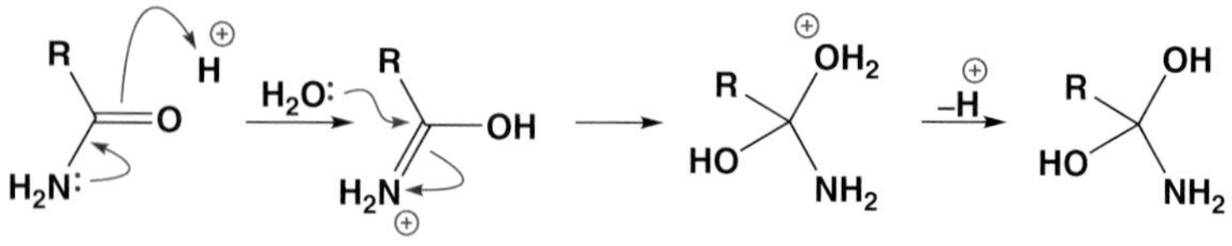

None of these substituents is now a good leaving group, but in acid solution one of them might be protonated. Which?

---

96. NH_2 is the most basic:

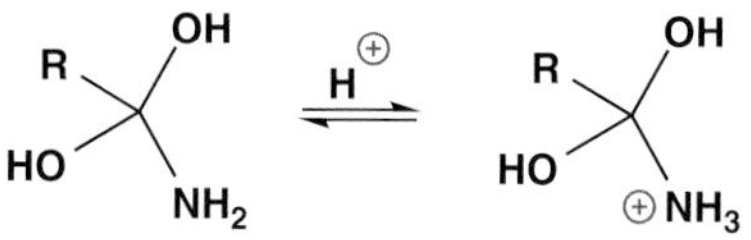

Now we have a good leaving group. Draw the next step.

97.

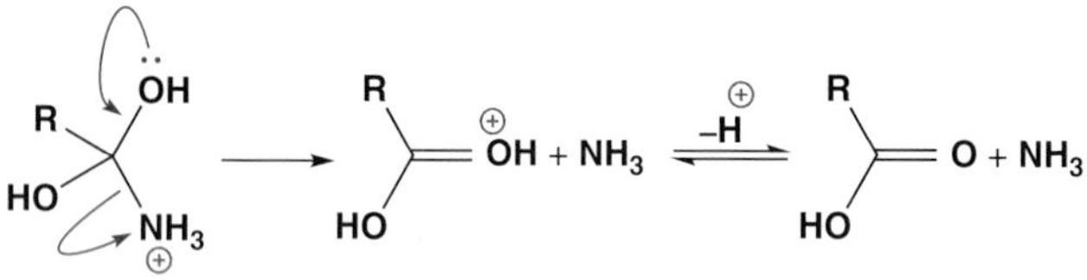

This reaction is very similar to some of the steps in acetal formation (*frames* **10–12**).

---

98.   So, to summarise, a base can catalyse the hydrolysis of esters or amides in two ways. State them.

99.

(1)   By driving over an unfavourable equilibrium using the mass-action effect.

(2)   By capturing the carboxylic acid product as an unreactive anion.

---

100.   Acid catalyses the same reaction also in two ways. State them.

101.

(1)   It catalyses the addition of the nucleophile by protonating the carbonyl group.

(2)   It turns what is otherwise a bad leaving group into a good one by protonation.

---

102.   This first catalytic function (*frame* **101** point (1)) can be carried out by electrophiles other than the proton. This applies particularly to carboxylic acids. The electrophile could attack either oxygen atom:

Which atom do you think will actually be attacked?

---

103.   The carbonyl oxygen atom will be attacked because the cation produced is delocalised over both oxygen atoms:

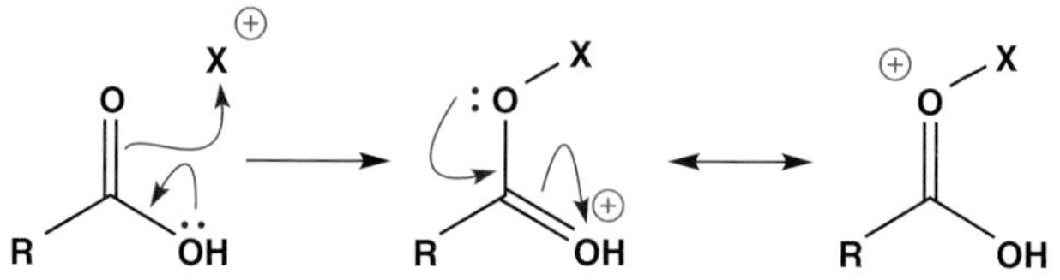

A good example of this kind of reaction is the attack of thionyl chloride on carboxylic acids. Initial attack occurs at carbonyl oxygen. Draw the products.

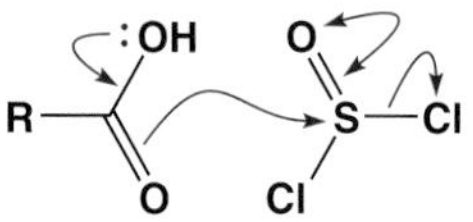

104.

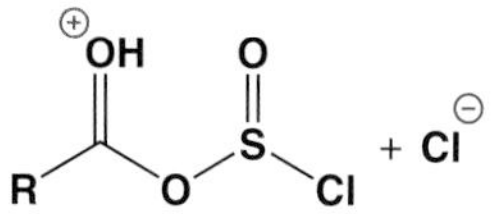

We now have within the system:

   (a)   A reactive carbonyl group. Why?
   (b)   A good leaving group. Which?
   (c)   A nucleophile. What?

105.

   (a)   Because it is protonated.
   (b) $ClSO_2^-$ which in fact decomposes to SO_2 and Cl^-.
   (c)   Chloride ion.

What do you think happens now?

106.

The net result is that a carboxylic acid has been converted into an acid chloride.

---

107. You may have noticed that this is the first example of a nucleophilic substitution at a carboxylic acid that we have seen, and we find in general that, unless we attack first with an electrophile, carboxylic acids are very unreactive towards nucleophilic substitution. Can you think of a reason for this?

---

108. There are really three related reasons: The carbonyl group is rather unreactive. The leaving group would have to be OH^-, notoriously one of the worst. Perhaps the most important of the three is that nucleophiles are bases and they therefore remove the acidic proton rather than attack the carbonyl group. How do we overcome these problems by reaction with thionyl chloride (*frames* **103–106**)?

---

109. The carbonyl group has become the reactive $C=O^+H$. We have a very good leaving group in $ClSO_2^-$. The removal of a proton has become an irrelevance since it's got to come off anyway and because of the electrophilic assistance we can use a very weak nucleophile, Cl^-. Other reagents which do the same job are $POCl_3$ and PCl_5.

---

110. How might we convert R(CO)OH to R(CO)OEt?

---

111.   We know that EtOH will displace Cl⁻ from the carbonyl group with base catalysis, so we need to make R(CO)Cl, and this we have just done. The whole scheme is:

$$R(CO)OH \xrightarrow{\text{SOCl}_2} R(CO)Cl \xrightarrow[\text{base}]{\text{EtOH}} R(CO)OEt$$

In fact we don't always need to go through R(CO)Cl, as acid catalysed reaction between the acid and EtOH often gives the ester in good yield. Mechanism?

---

112.   Protonation gives the most delocalised cation (*frame* **95**):

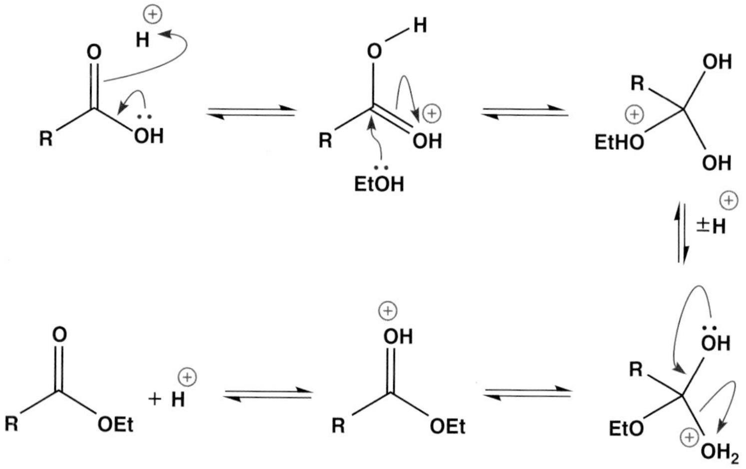

This is another example of an equilibrium, so we make an ethyl ester from R(CO)OH, EtOH and acid in solution ____ _____, and we hydrolyse an ester in ____ _____ with acid catalysis.

---

113.   We make the ester in *ethanol* and hydrolyse it in *water*. Another useful compound is the anhydride R(CO)O(CO)R. How could we make that from R(CO)OH using reactions we've discussed?

---

114.   Again we need to displace Cl⁻ by R(CO)O⁻:

This is practically the end of the second section, so here are some general review questions.

---

115.   Arrange these compounds in order of reactivity towards water; all reactions are to give the carboxylic acid:

$$CH_3(CO)O(CO)CH_3 \qquad CH_3(CO)NH_2$$
$$CH_3(CO)Cl \qquad CH_3(CO)OEt$$

---

116.   Chloride > anhydride > ester > amide. In fact, the chloride explodes with cold water, the anhydride reacts with cold water, the ester reacts with dilute acid or base but the amide only hydrolyses with boiling 70% acid or 10% caustic soda. Ask your adviser if you are in doubt about this.

---

117. If you hydrolysed an ester labelled with heavy oxygen, in acid or base, would the ^{18}O end up in the acetic acid or the ethanol?

---

118. In both acid and base it would end up in the ethanol; this is one of the pieces of evidence used to establish the mechanisms we have been discussing. If you want to read more about this see Clayden[1] or Klein[2].

---

119. How would you carry out this multi-step synthesis?

---

[1]J. Clayden, N. Greeves and S. Warren. *Organic Chemistry*. 2nd Ed. OUP, (2012). ISBN: 978-0-19-927029-3, page 201, 211.
[2]D. Klein. *Organic Chemistry*. 2nd Ed. Wiley, (2015). ISBN: 978-1-118-45228-8, page 1016.

120.

There are other good routes: if you have one discuss it with your adviser.

This is the end of the second part of the program.

CHAPTER 3

NUCLEOPHILIC SUBSTITUTION AT THE CARBONYL GROUP WITH COMPLETE REMOVAL OF CARBONYL OXYGEN

Contents

Chemistry of the Carbonyl Group:
A step-by-step approach to understanding Organic Reaction Mechanisms,
Revised Edition. Timothy K. Dickens and Stuart Warren.
© 2018 John Wiley & Sons Ltd. Published 2018 by John Wiley & Sons Ltd.

Concepts assumed:

- Rigidity of olefin two-dimensional structure.
- Geometrical isomerism. Mechanism of the E2 reaction.

Concepts introduced:

- Absence of acid chloride-like substitution in aldehydes and ketones.
- Possibility of loss of carbonyl oxygen from tetrahedral intermediate.
- Geometrical isomerism of oximes.
- Usefulness of different reagents using contrasting conditions for the same synthetic step.
- Non-nucleophilicity of amide nitrogen atoms.
- Use of high-boiling solvents.
- Medium-ring compounds.
- Characterising a compound as a stable crystalline derivative.

Concepts reinforced:

- Instability of compounds containing two atoms, both with lone pairs, bonded to the same carbon atom.
- Incompatibility of strongly basic nucleophile and strongly acidic conditions.
- Electrophilic attack of S and P compounds on carbonyl oxygen.
- Electrophilic substitution in the benzene ring.

121.   You may have the impression from the last section that aldehydes and ketones can't do substitution reactions, but this isn't true.
We don't of course get reactions like this:

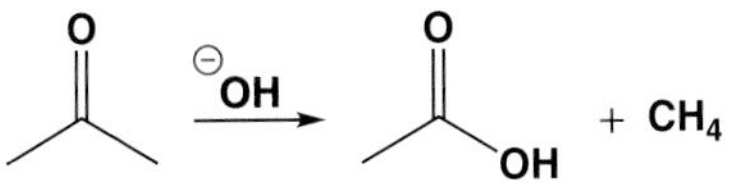

Why not?

---

122.   The mechanism would have to be:

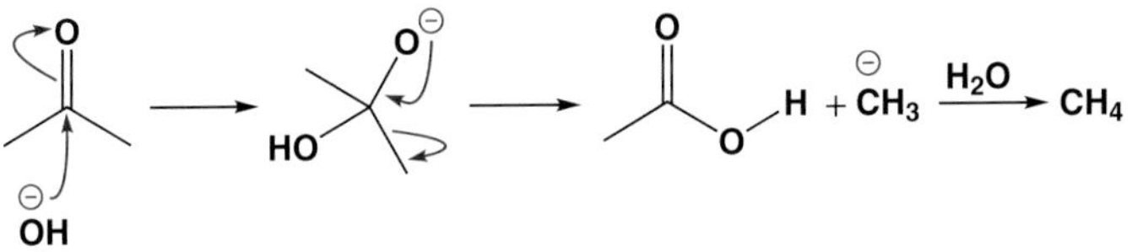

and $CH_3{}^-$ is far too basic to be displaced. We are going to look at a
new kind of substitution reaction. Draw out the addition of aniline,
$PhNH_2$, to acetone to give a neutral adduct.

123.

If this addition reaction is to be extended into a substitution, we must find a leaving group. Neither CH_3, OH, nor PhNH are good leaving groups but either OH or PhNH can *become* so. How?

---

124.   By protonation (as in amide hydrolysis, *frames* **94–97**). If PhNH is protonated and eliminated, we just reverse the reaction back to the starting materials, but see what happens if you protonate and eliminate OH. Draw the reaction.

125.

What you have done is to make the carbonyl oxygen atom the leaving group: the reactions we are going to explore in this section all involve complete removal of carbonyl oxygen during a substitution reaction.

---

126.   The product of this last reaction is an imine, containing a C=N double bond. The formation of these compounds is an equilibrium and they are very easily hydrolysed. More stable imines are formed from hydroxyl-amine, HONH$_2$. Write down the complete reaction between this compound and Ph(CO)Me (using the N atom as the nucleophile).

127.

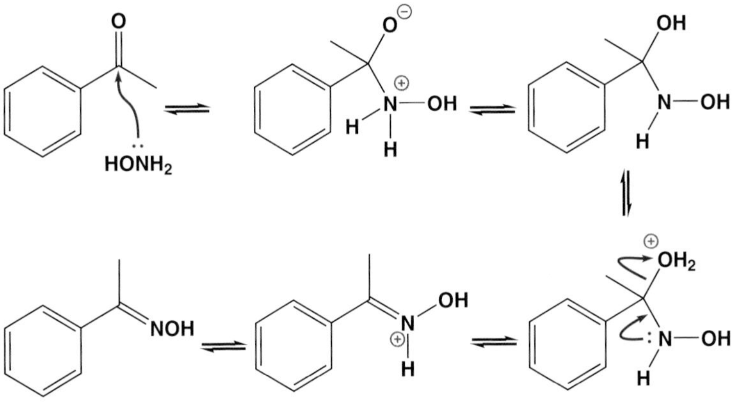

Draw out the full (two-dimensional) structure of this product.

128.

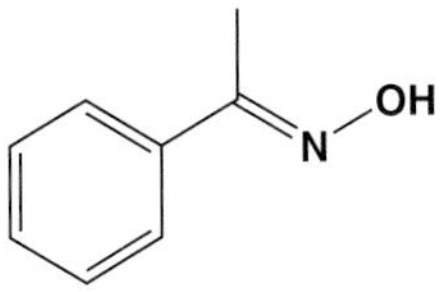

How many isomers of this compound are there?

---

129. There are in fact two: the C=N bond is just as rigid as the C=C double bond, and these compounds, oximes, are like olefins with one substituent missing; there are *cis* and *trans* forms:

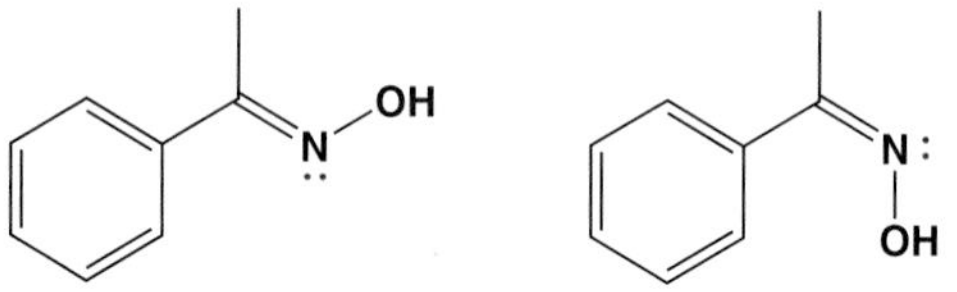

130. Similar reactions occur with other amines, particularly hydrazine, NH_2NH_2, and its derivatives which form good, stable, crystalline imines known as hydrazones. What compound would be formed from hydrazine and acetone?

131.

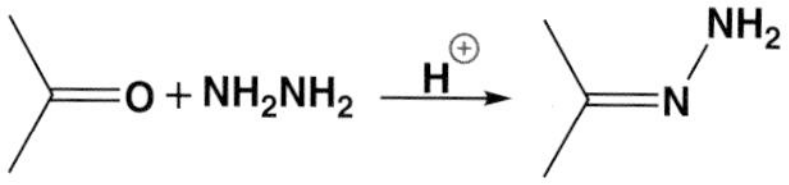

Supposing there were more acetone around what might happen now?

132. The other end of the molecule could react too:

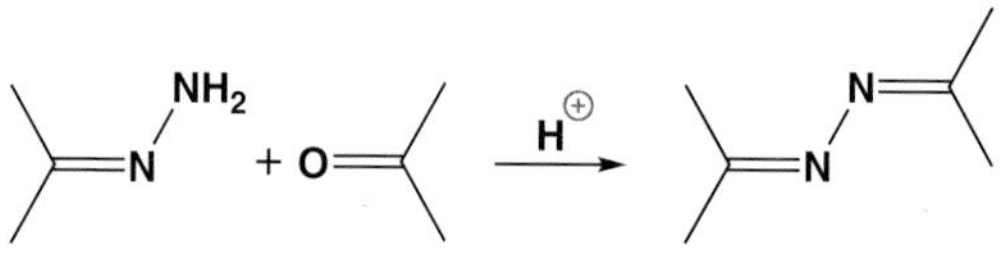

133. Semicarbazide, $NH_2NH(CO)NH_2$, also reacts well: what product would be formed here with acetaldehyde?

134.

This nitrogen atom is in fact the only one to react. Why don't the other two react as well?

---

135. The other two nitrogen atoms have their lone pairs conjugated to the carbonyl group: they are in fact amide-like. It is only the terminal nitrogen atom that is fully nucleophilic.

---

136. We have met the hydride ion in the guise of $NaBH_4$ and $LiAlH_4$ twice already (*frames* **41–45** and *frames* **76–79**). If we now use hydride ion to remove the carbonyl oxygen atom altogether, we shall obviously get a hydrocarbon:

$$R_2C = O \longrightarrow R_2CH_2$$

What could be an intermediate in this reaction be?

---

137.   The obvious intermediate is found by adding a hydride ion to the carbonyl compound in the usual way:

Under strongly acidic conditions this alcohol could give a carbonium ion. Draw this.

138.

All we have to do now is to add a hydride ion to the carbonium ion to get the hydrocarbon. You may have been nursing a growing feeling that all is not well with this idea, and you are right. What is wrong?

---

139.   If you're not sure, consider whether there isn't something incompatible with the nature of the hydride ion, or $NaBH_4$ or $LiAlH_4$, and the conditions we have outlined for this reaction.

---

140. We need strong acid to protonate and eliminate the carbonyl oxygen atom, and we can't possibly use any of the sources of hydride ion in acid solution: $NaBH_4$ would react violently to give hydrogen gas, and $LiAlH_4$ or NaH would explode. We therefore use a dissolving metal reduction in strong acid. This reaction, the Clemmensen reduction, may use the principle we have outlined here, but its mechanism is unknown in detail.

$$R_2C = O + Zn/Hg \xrightarrow{\text{Conc. } H^+} R_2CH_2$$

---

141. An alternative method begins with the formation of a hydrazone. We shall use cyclohexanone here for a change. Draw the product formed from this ketone and hydrazine (the hydrazone).

142.

The hydrazone has all the elements of the product we want, cyclohexane, plus two nitrogen atoms. All (!) we have to do is to move two hydrogen atoms from nitrogen to carbon. This can be done in a very strong base:

Complete the reaction mechanism to give cyclohexane.

143.

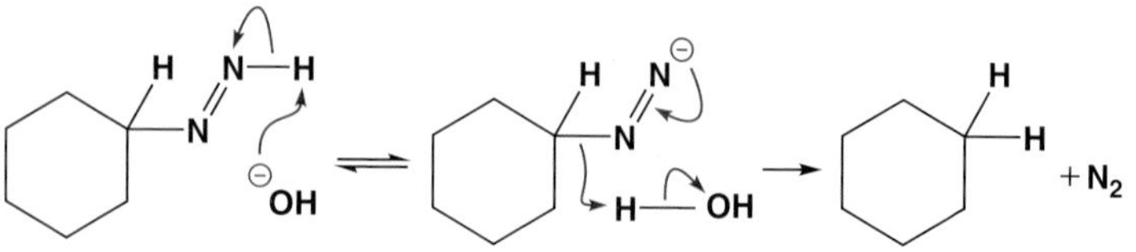

The reaction is usually done at high temperatures in ethylene glycol, a high-boiling polar solvent, and is called the Wolff–Kischner reduction:

$$R_2C = O + NH_2NH_2 + NaOH \xrightarrow[200\ °C]{\text{Reflux in glycol}} R_2CH_2 + N_2$$

---

144. Yet another method is to make the dithioacetal from the ketone, say $Ph(CO)CH_2CH_3$, and $HSCH_2CH_2SH$. Draw this.

---

145. If you can't do this, look back at the *frames* on cyclic acetal formation (*frames* **62** and **63**): it's really the same reaction.

146.

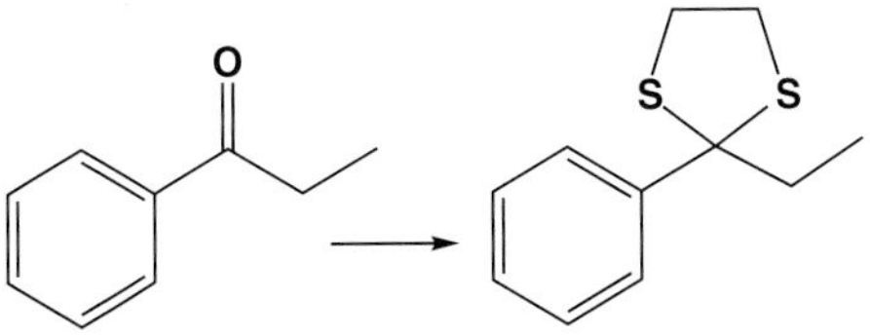

The dithioketal can be reduced directly to the hydrocarbon by hydrogenation over the sulphur-removing catalyst Raney nickel:

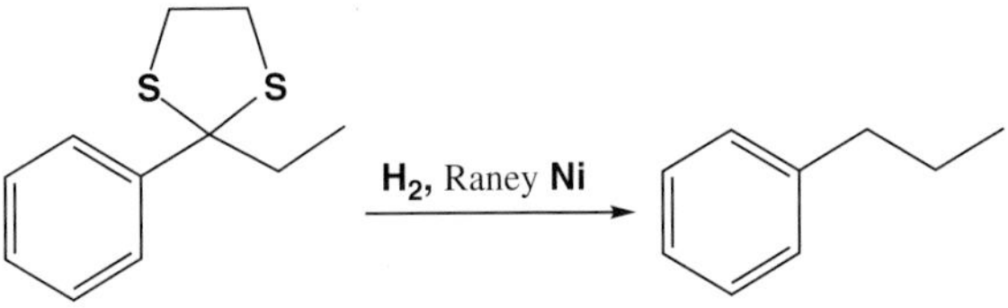

---

147. We now have methods of reducing ketones to hydrocarbons in acidic, alkaline and neutral solutions. This is useful since we may have molecules that are sensitive to some of these conditions. Which methods would you use for these reactions?

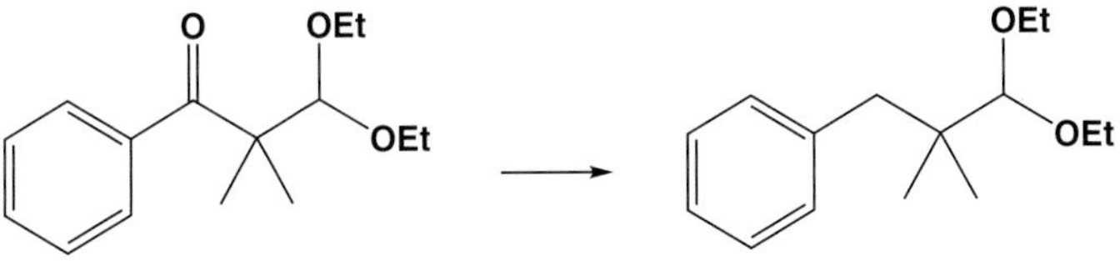

(a)

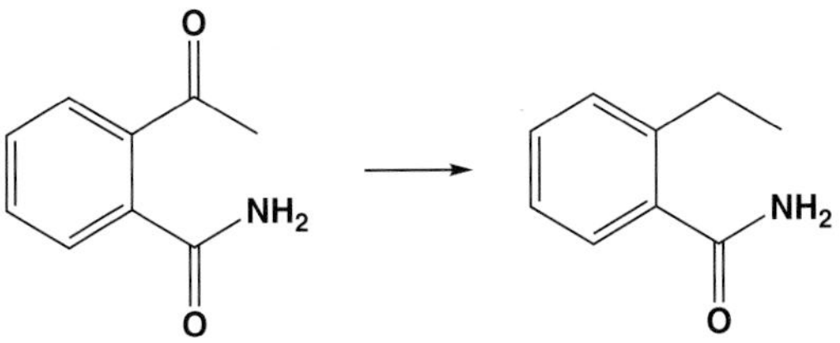

(b)

148.

(a)   Wolff–Kischner. Clemmensen would destroy the acetal.
(b)   Dithioketal. Either of the other methods would do nasty things to the amide.

This is a good place to stop if you'd like a rest.

---

149.   The idea of adding a nucleophile to a carbonyl compound to give an alcohol and then dehydrating the alcohol to a carbonium ion in acid solution, and finally adding another nucleophile to the carbonium ion, could be somewhat extended. In *frames* **137** and **138**, we discussed this idea for H⁻ as nucleophile, but it didn't work out too well. Draw the sequence out for Cl⁻ as nucleophile on a general ketone, $R_2C{=}O$.

150.

It turns out that we can't do this with HCl; we need a stronger electrophilic catalyst than the proton for the first step. We use instead PCl_5, which reacts by ionising to Cl⁻ and PCl_4^+. The products are R_2CCl_2 and $POCl_3$. See if you can complete the mechanism.

151.

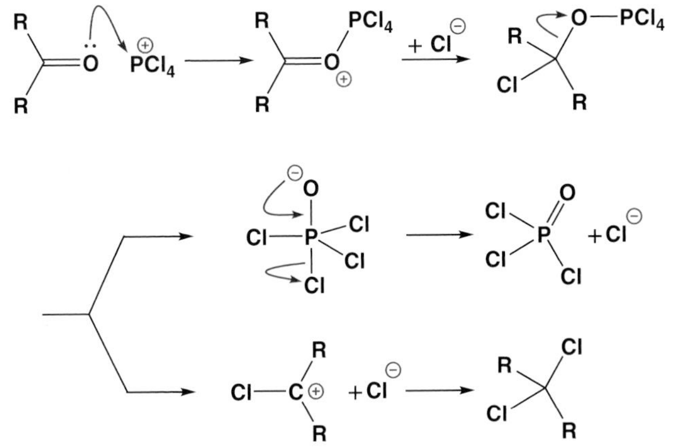

Draw the product you would get from PCl_5 on $PhCOCH_2Ph$.

152.

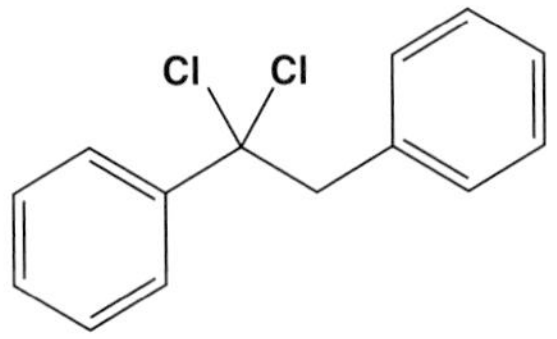

Supposing you now reacted this with EtO^- in EtOH. What would you get?

153. Stuck? Ethoxide ion is very basic and likes to remove protons, even protons attached to carbon atoms. We also have good leaving groups in the molecule and that combination looks like making an elimination reaction …

154.

A double elimination by the E2 mechanism.

---

155. At this point, you have a choice. We want to deal with some substitutions on the benzene ring using mechanisms like that for the nitration of benzene. If you're quite familiar, read on. If you want to brush up on the subject, consult a standard text on substitution reactions of aromatic molecules, see Clayden[1] or Klein[2]. If you know nothing about the subject, either skip to *frame* **165** or read Clayden or Klein and do your best.

---

[1] J. Clayden, N. Greeves and S. Warren. *Organic Chemistry.* 2nd Ed. OUP, (2012). ISBN: 978-0-19-927029-3, chapter 16.
[2] D. Klein. *Organic Chemistry.* 2nd Ed. Wiley, (2015). ISBN: 978-1-118-45228-8, chapter 19.

156. With very reactive carbonyl compounds, such as chloral $Cl_3C(CO)H$, we can even add aromatic compounds in strong acid. Draw a mechanism for the first step, the formation of the alcohol:

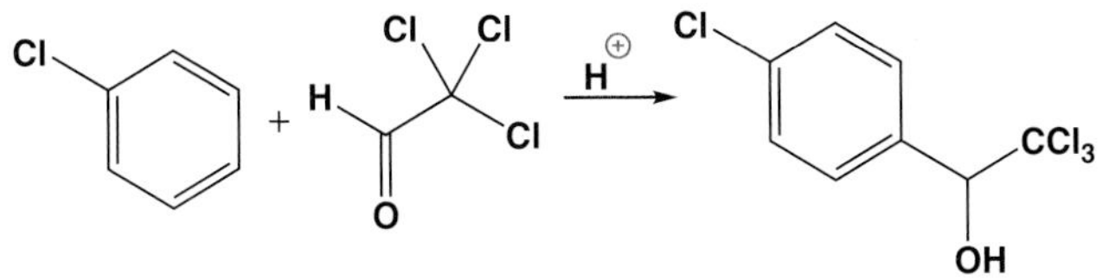

157.

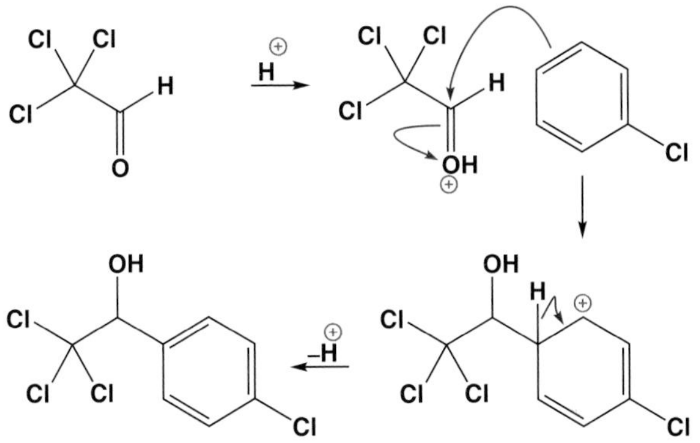

Now a second molecule adds on:

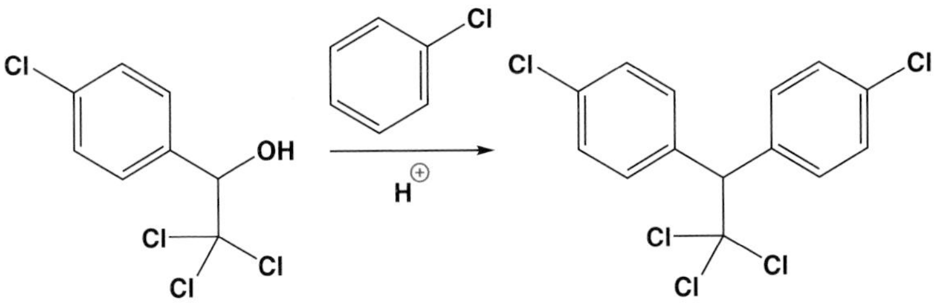

Suggest a mechanism for this step.

158.

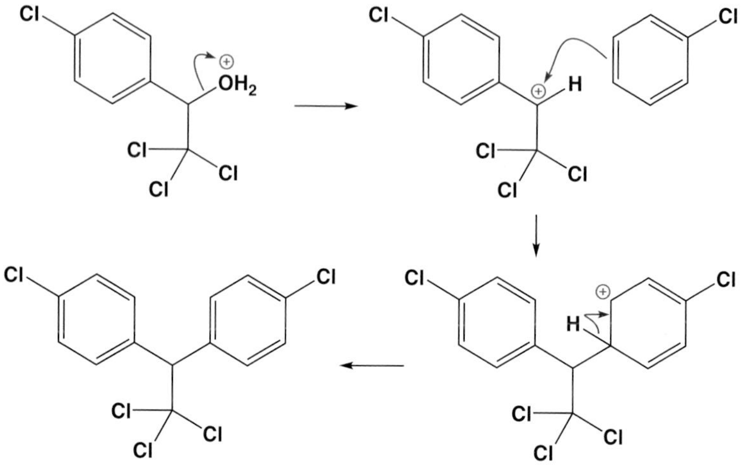

You may recognise this product as DDT, the once famous, now infamous, insecticide. Its use is now controversial because of the build-up of organic chlorocompounds in animals throughout the world.

---

159. Another reactive carbonyl compound is formaldehyde. Draw out the first step of its addition to benzene in acid solution.

160.

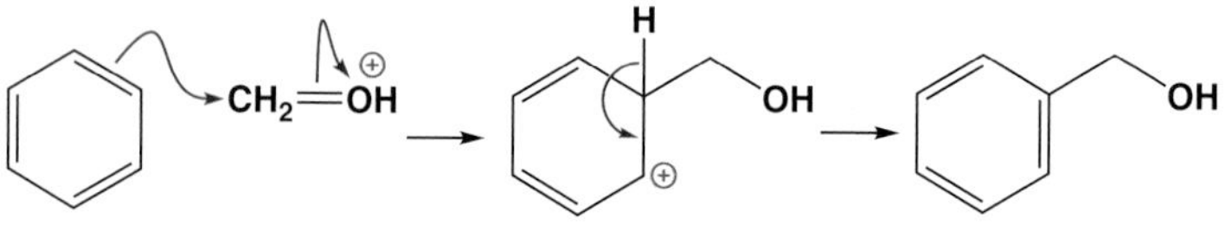

If HCl is used as the acid catalyst, Cl⁻ instead of the aromatic ring becomes the nucleophile for the second step. Draw this.

161.

This is then a general reaction for adding a $ClCH_2$ group to an aromatic ring known as chloromethylation. The reaction is carried out in a single step, CH_2O and HCl being added to the aromatic compound.

---

162. The catalyst used for the DDT synthesis is H_2SO_4 and that for chloromethylation is HCl. In the DDT synthesis, we added two molecules of aromatic hydrocarbon, but in chloromethylation the second step had Cl^- as nucleophile. Comment?

---

163. We deliberately used H_2SO_4, an acid with a non-nucleophilic anion, in the DDT synthesis, and HCl, an acid with a nucleophilic anion, in chloromethylation. We also used H_2SO_4 for the same reason in the dehydration of tertiary alcohols produced by the addition of Grignard reagents to ketones. Draw out an example of this reaction.

164.

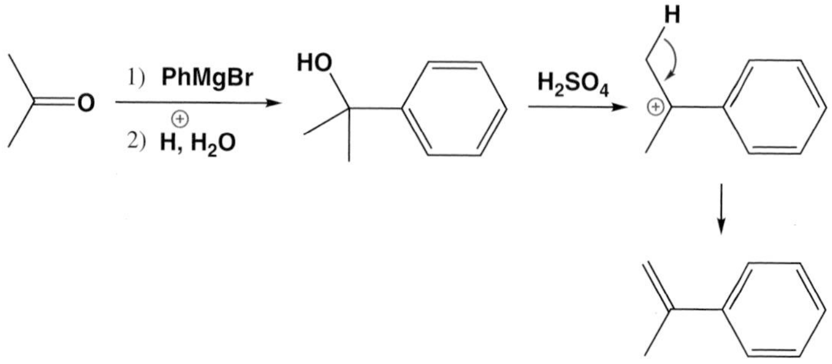

This is then our final example of substitution removing the carbonyl oxygen atom completely: here you see C=O being replaced by C=C.

---

165. At the end of this section, there are two review questions: aldehydes and ketones are often characterised as 2,4-dinitrophenylhydrazones, as these are usually highly crystalline orange compounds. Draw the formation of this derivative from benzaldehyde giving reagent, catalyst and the structure of the product.

166.

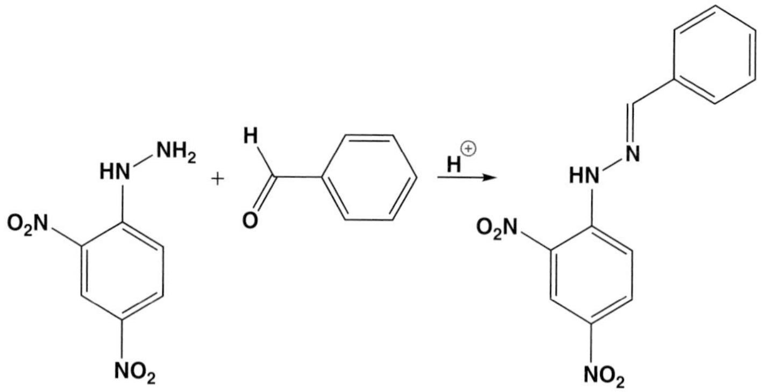

167.   The smallest stable cyclic acetylene is in the nine-membered ring. If you had a sample of a nine-membered ring ketone, how would you attempt to make the cyclic acetylene from it?

168.

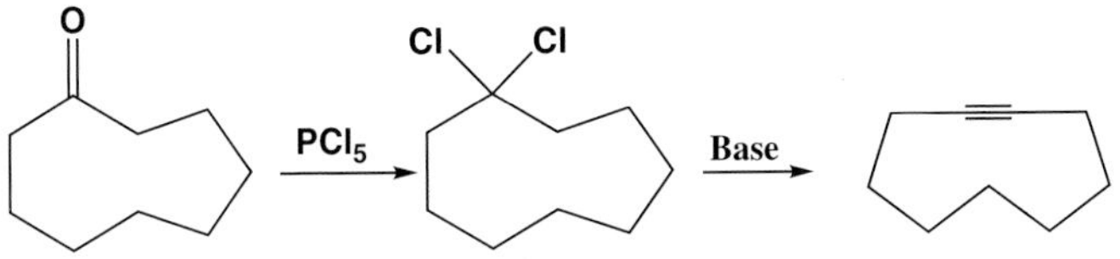

This is the end of the third section of the program.

CHAPTER 4

CARBANIONS AND ENOLISATION

Contents

Chemistry of the Carbonyl Group:
A step-by-step approach to understanding Organic Reaction Mechanisms,
Revised Edition. Timothy K. Dickens and Stuart Warren.
© 2018 John Wiley & Sons Ltd. Published 2018 by John Wiley & Sons Ltd.

Concepts assumed:

- Hydrogen bonding.
- Racemisation.
- Polymerisation.

Concepts introduced:

- Simple carbanions don't occur.
- Stability of enolate anions.
- Tautomerism.
- Reactivity of enols and enolates at carbon.
- Selectivity in reactions by:
  - choice of acid or base catalyst;
  - choice of metal in organo-metallic reagent.
- Subtle arguments in rationalising difference between acid and base catalysed halogenation of ketones.

Concepts reinforced:

- Effects of equilibria. Drawing transition states.
- Effects of substituents on stability/reactivity.
- Relationship of nucleophilic addition, substitution and enolisation.
- Synthesis of acid chlorides.
- Use of organo-metallic reagents in synthesis.
- General relation of synthesis and mechanism.

169.   Carbonium ions are familiar intermediates in many simple reactions (S_N1, E1, etc.), and we have met many of them in this program. Carbanions are another matter. Though there are some exceptions, such as Br_3C^-, it is a good general rule to say that simple carbanions do not occur as reactive intermediates. If we want to remove a proton from a carbon atom to make an anion, we need somewhere to park the negative charge and there is nothing better for this job than the carbonyl group. Draw arrows to show the formation of an anion from acetone:

170.

This anion, often loosely called a carbanion, is delocalised with the charge shared between the oxygen and carbon atoms. Draw arrows to show this.

171.

Now show how this anion can react with a proton on carbon and on oxygen to give two different products.

172.

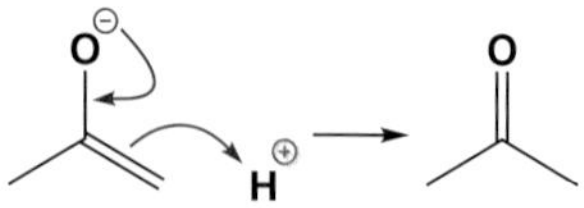

(172a)

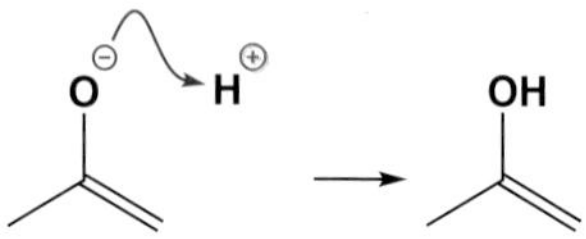

(172b)

These two compounds have identical structures except for the position of one proton: this is clearly a special case of isomerism and it is called tautomerism. **(172a)** is called the keto and **(172b)** the enol form of acetone. Draw out the mechanism for the conversion of the keto to the enol form in base.

173.

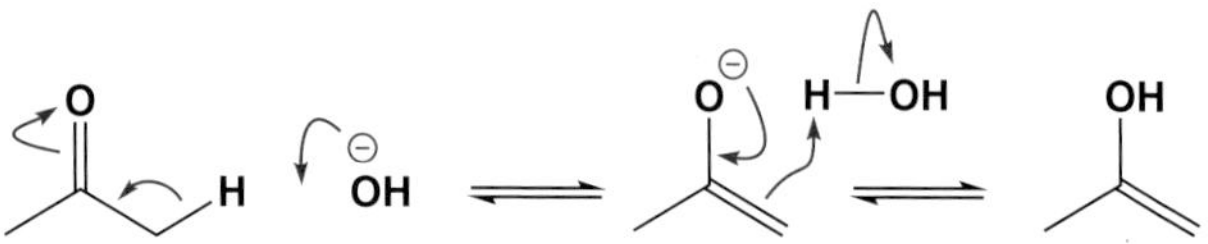

This is an equilibrium, catalysed by acid as well as by base. Show how the enol of acetaldehyde could be formed with acid catalysis.

174.

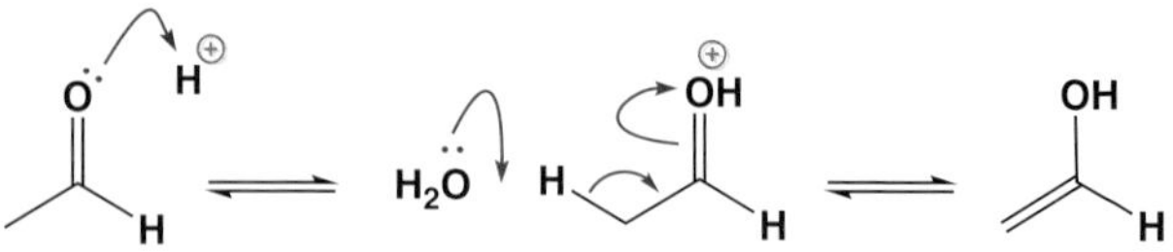

Now draw the enol form of this ketone: $CH_3(CO)CH_2(CO)CH_3$.

---

175. There are in fact two possible enol forms:

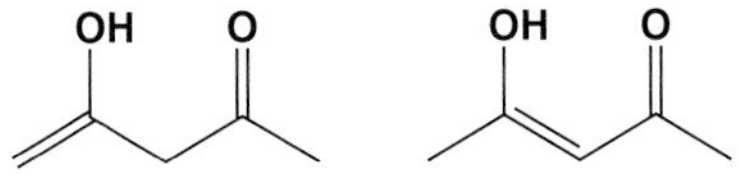

Which is the more stable?

---

176. The second: not only because the double bond is conjugated with the carbonyl group but because of intramolecular hydrogen bonding:

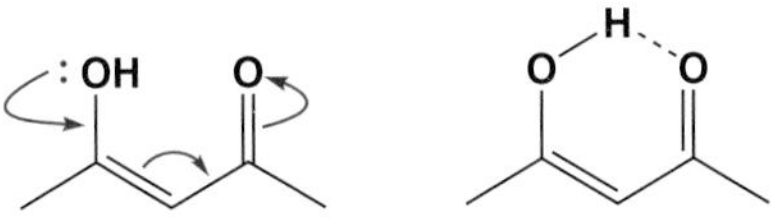

This compound in fact exists almost totally as the enol under normal conditions, in contrast to acetone that is entirely in the keto form. Many other carbonyl compounds are mixtures of the two.

---

177. The 'carbanions' we formed using the carbonyl group are like enols and are called enolate anions. Draw the enolate anion from this ketone:

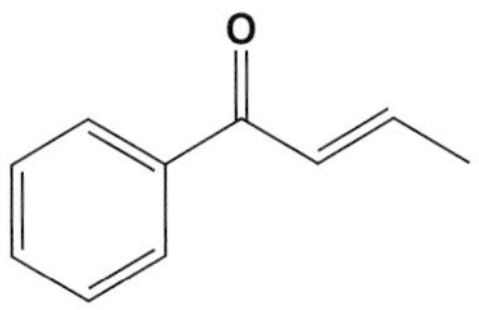

178.   This time we have to remove a rather distant proton, but the charge still gets to the carbonyl group:

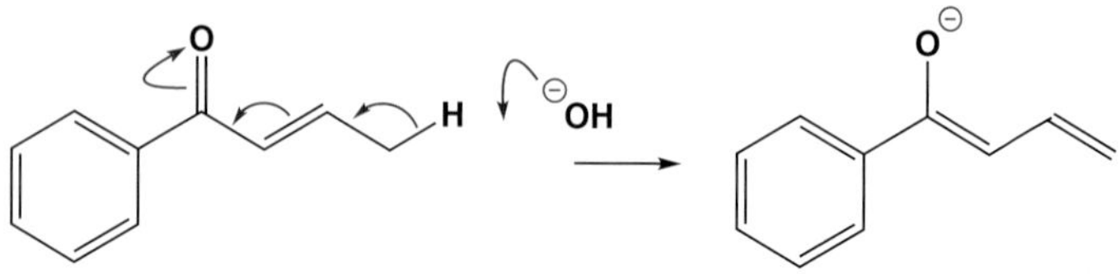

Draw the arrows for the re-protonation of this enolate to give the original compound.

179.

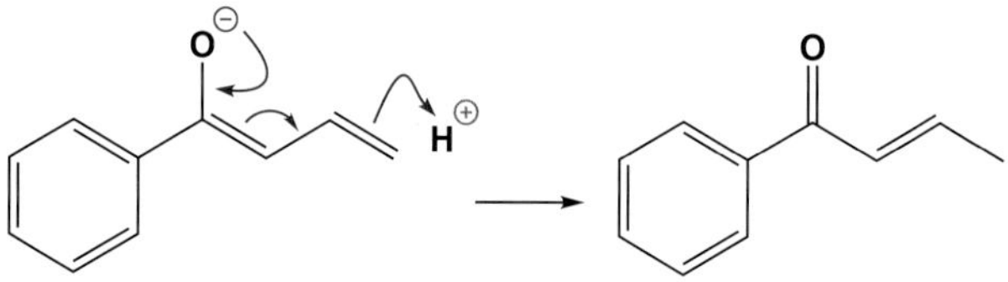

Now form the enolate anion from this ketone:

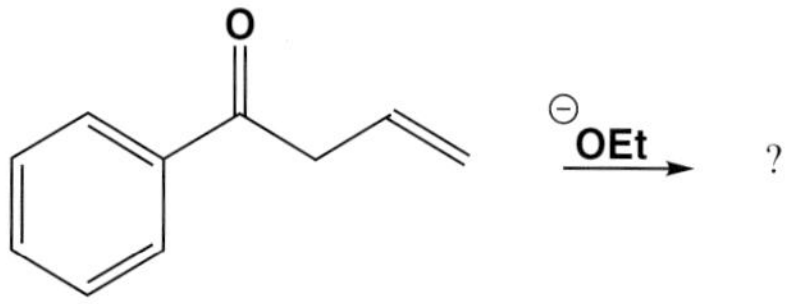

180. You should be adept at this by now:

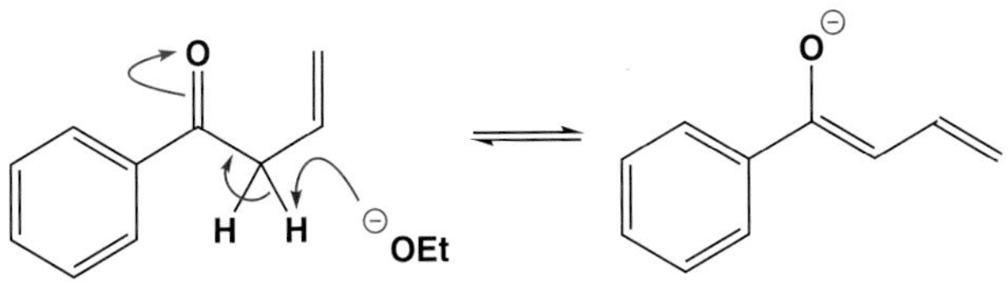

Now put a proton back to give a ketone.

---

181. There's a bit of a catch here. Did you notice that the enolate anions formed in *frames* **178** and **180** are the same? Perhaps you want to think again.

---

182. If the two enolates are the same, then they must protonate to give the same ketone. Why in fact do we form this one and not the alternative **(182b)**?

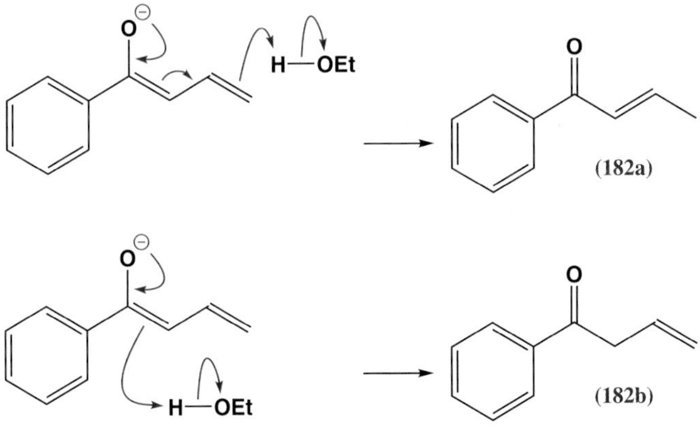

(182a)

(182b)

183. Because the whole thing is an equilibrium and the conjugated ketone **(182a)** is the more stable. What happens then if we dissolve **(182b)** in ethanol containing a small amount of EtO⁻?

---

184. Small amounts of the enolate ion will be formed that will re-protonate to give **(184a)**. Ketone **(184b)** is therefore quickly transformed into **(184a)**.

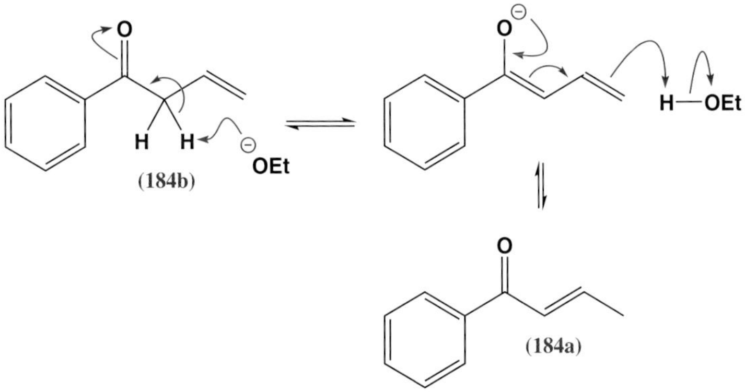

(184b)

(184a)

---

185. What happens if optically active **(185)** is dissolved in ethanol containing a catalytic amount of ethoxide ion?

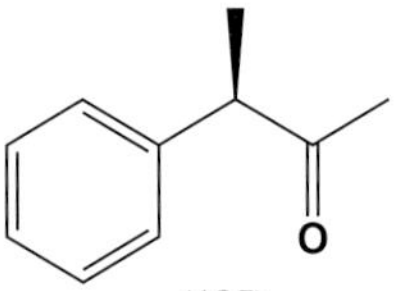

(185)

---

186.   Again, an equilibrium concentration of the enolate anion is formed, and when it is re-protonated it must give racemic product since it is planar and the proton can add on to either side:

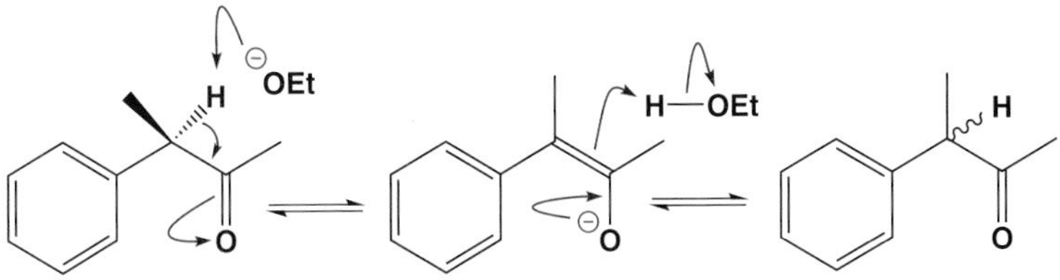

187.   You should be getting the general idea by now: one more example of this kind of thing will do. What happens if ketone **(187)** is dissolved under the same conditions?

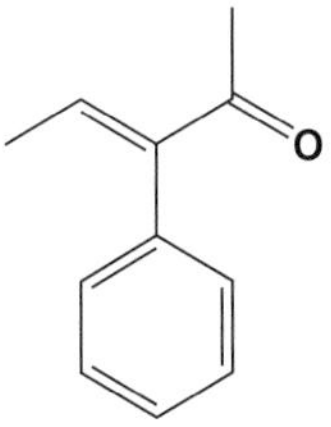

(187)

188. This time the enolate anion can rotate about what was a double bond in the original compound, giving a mixture of *cis* and *trans* forms.

This is a good place to stop if you want a rest.

---

189. So far, we have reacted enolate ions with protons. They also react with many electrophiles in just the same way: the electrophile attacking at carbon. Make the enolate ion from Ph(CO)Me and attack it with bromine.

---

190. If you're in trouble, bring the negative charge back from the oxygen atom, just as you have been doing so far, but this time attack one end of the bromine molecule with it, expelling the other bromine atom as Br⁻.

191.

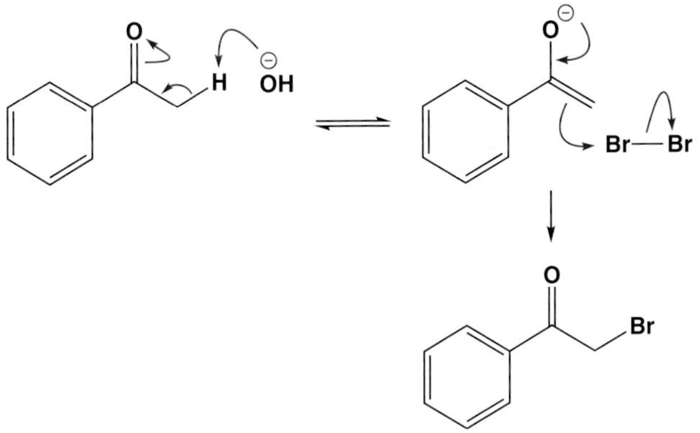

With base still present in the reaction mixture, would anything happen to this product?

---

192. Bromide ion could be displaced, but in fact it is not. Have you noticed that there are two more protons waiting to be removed?

---

193. The process is repeated with another proton:

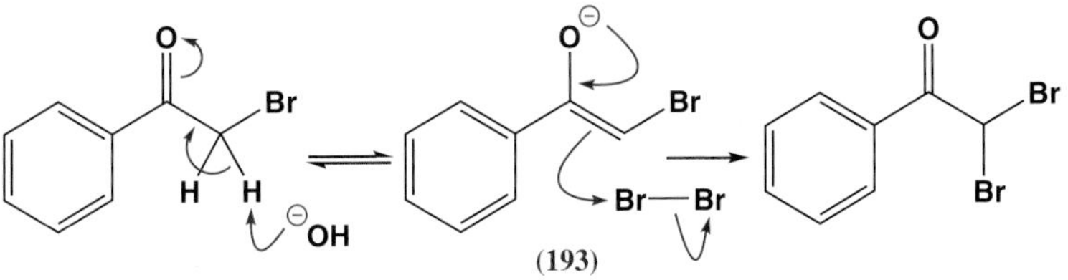

(193)

Will this second enolate ion **(193)** be formed more or less easily than the first (in *frame* **191**)?

---

194.   *More* easily, because the inductive effect of the bromine atom makes the hydrogen easier to remove. We can't therefore stop this reaction at the *mono*-bromo stage, nor at the *di*-bromo stage: what happens next?

---

195.   The process is repeated for a third time:

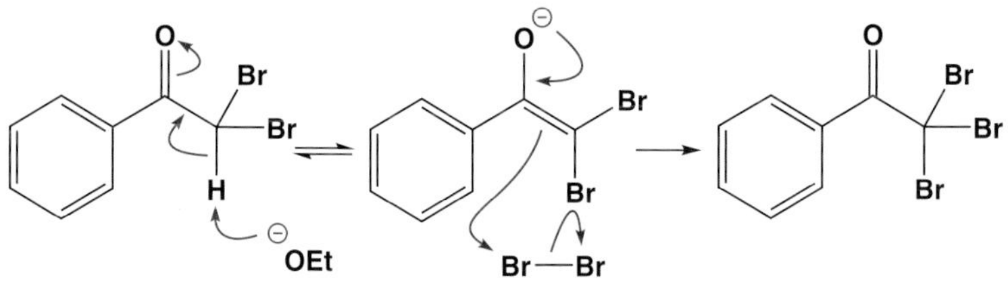

In the presence of enough bromine, therefore, the ketone is converted rapidly into the tri-bromo ketone. But the reaction doesn't stop, even here. Do you remember what we singled out as one of the rare stable simple carbanions? How could another molecule of hydroxide ion attack the product to give this anion?

---

196. In a straightforward substitution reaction:

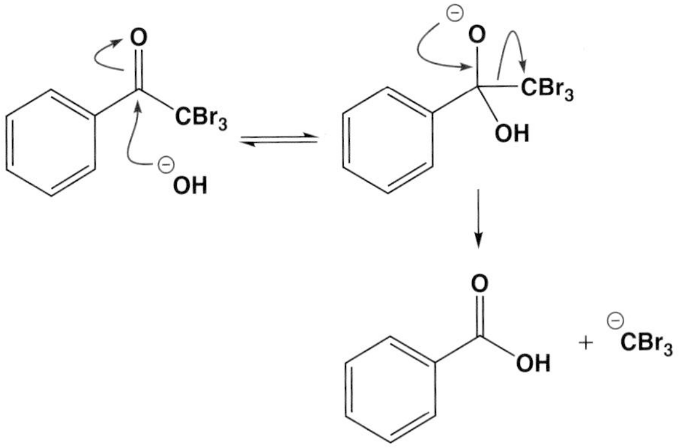

The carbanion, though *relatively* stable compared to other carbanions, is still reactive enough to pick up a proton from water to give $CHBr_3$. Write down a summary equation for this whole process, starting from PhCOMe.

197.

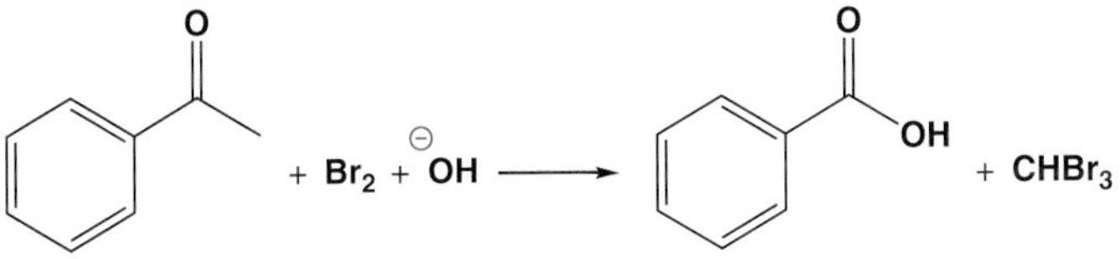

This reaction works equally well with chlorine or iodine and is known as the chloroform or iodoform reactions in these cases, after the names of the products.

---

198. We do, of course, need a way of making the *mono*-bromo derivative of a ketone, and it turns out that we can do this by halogenation in acid solution. Take the same ketone, PhCOMe, make the enol in acid solution and react it with bromine.

---

199.   Congratulations if you got this right all in one go!

200.   Now we come to the slightly tricky question of why this reaction can be stopped at this stage. Let's look at the enolisation step (marked ★ in *frame* **199**). Draw the transition state for this step. If you've forgotten how to draw transition states, look at *frames* **54 –58**.

201.

We want to compare this transition state with the one for the same step with the *mono*-bromo compound, so we have drawn them side by side. Note that, in these transition states, a positive charge is smeared out over all the atoms connected by the dotted line. Therefore, any substituent like Br that is electron withdrawing will destabilise the transition state. The first step to give the *mono*-bromo compound goes more quickly than the step to give the *di*-bromo compound, and we can stop the reaction by using only so much bromine at the first stage.

202. We have made acid chlorides in the program already (*frames* **103 –106**) by the reaction between $SOCl_2$ and $R(CO)OH$. PCl_5 and PBr_5 do much the same thing:

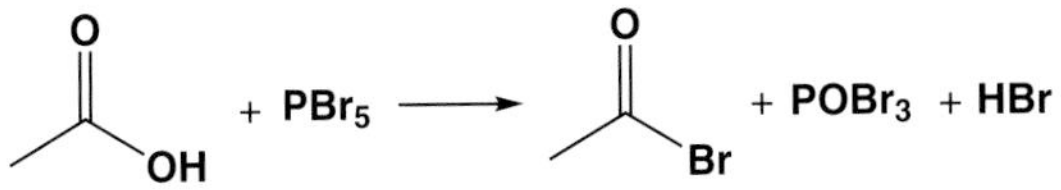

What further reaction might happen under these conditions if bromine were also present in the reaction mixture?

---

203. A clue is in the presence of HBr, which could catalyse the enolisation of the acid bromide.

---

204. The acid bromide enolises and reacts with bromine:

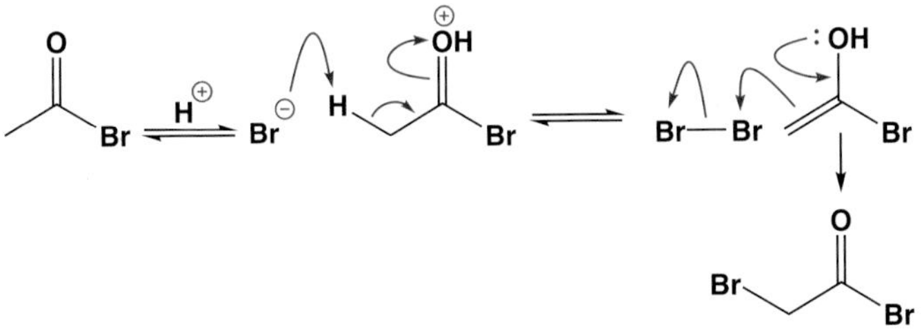

Since this is an acid-catalysed reaction, it can be stopped at this stage. This valuable reaction, known as the Hell–Volhard–Zelinsky (HVZ) reaction, is normally carried out by treating the acid with red phosphorus and bromine. PBr_5 is thus made on the spot.

---

205.   What would be the product in this reaction?

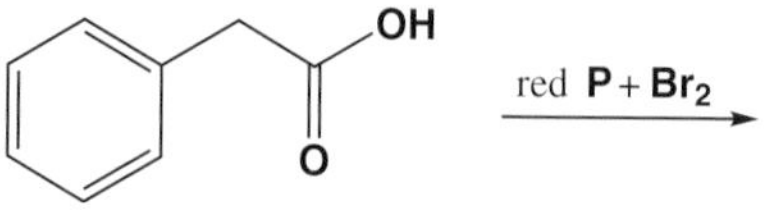

red **P** + **Br₂**

206.

red **P** + **Br₂**

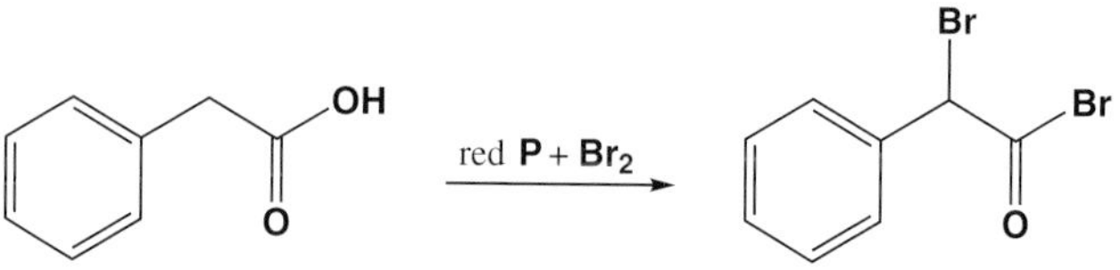

And what is the product in this reaction?

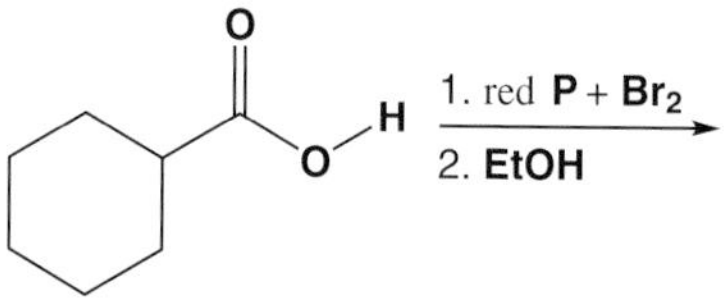

1. red **P** + **Br₂**
2. **EtOH**

207.

red **P** + **Br₂**

EtOH

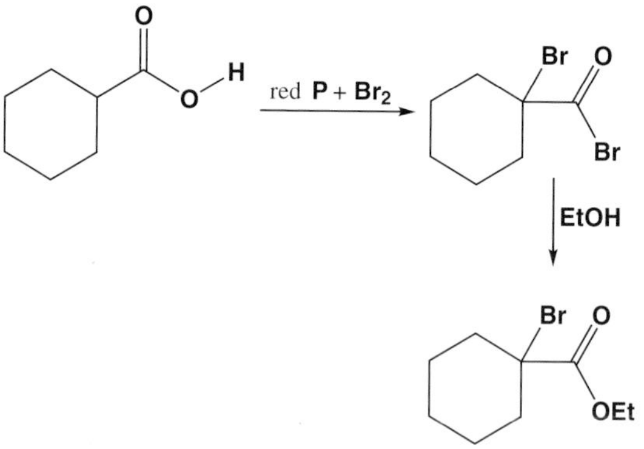

We don't usually try to make Grignard reagents from α-bromo esters like this compound. Why not?

208. Because Grignard reagents react with esters, and the Grignard reagent from this compound would therefore react with itself, that is polymerise. We use the zinc compound instead as these are less reactive and don't polymerise. We make them in the same way as Grignard reagents, and they react in the same way with ketones. What would happen here:

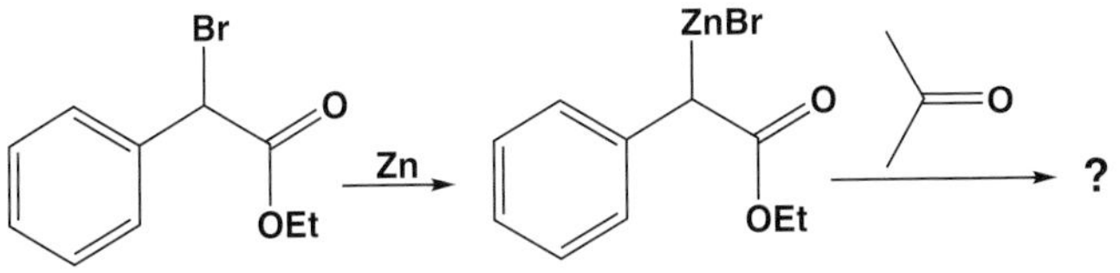

209.

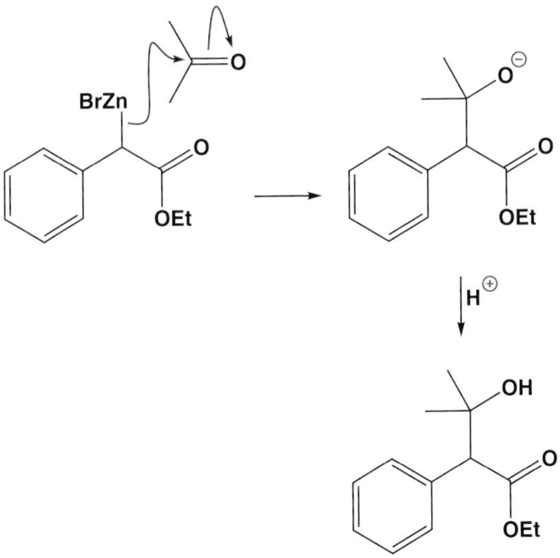

What do you think happens to this during the acidic workup?

210.   It dehydrates spontaneously to give an α–β unsaturated ester. We have already met this reaction in *frame* **40**.

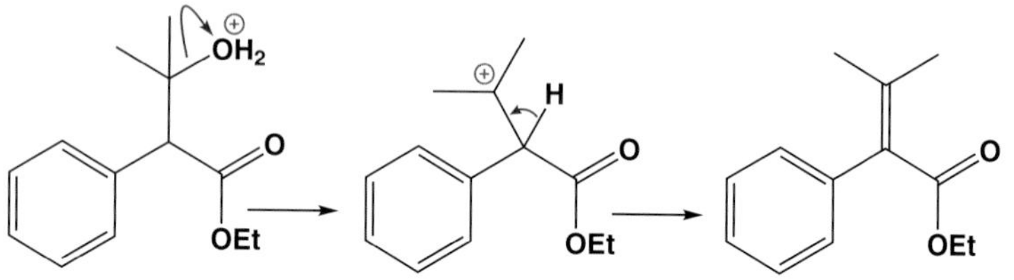

---

211.   We said that these reactions are valuable so let's prove it. How would you carry out this synthesis?

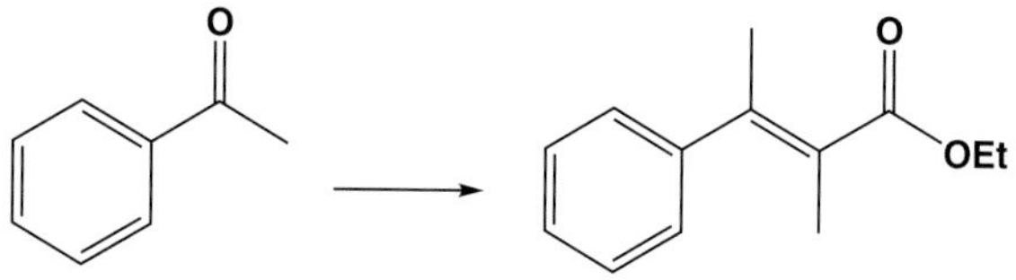

212.

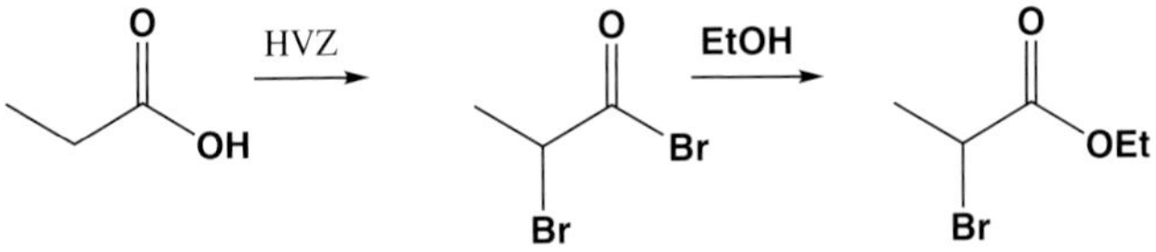

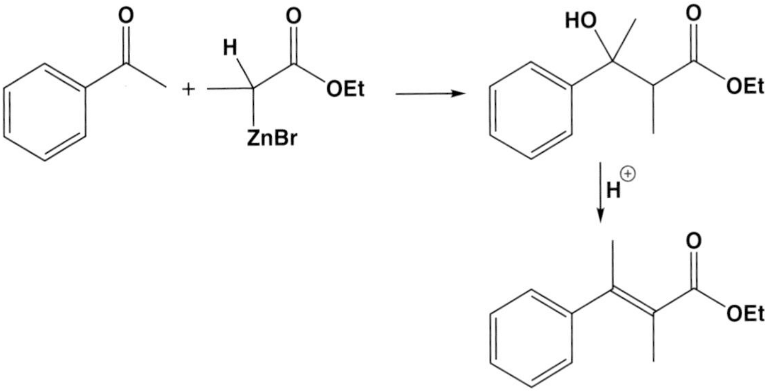

---

213. In the next section, we shall be looking at more reactions of enols which make new carbon–carbon bonds. We'll finish off this section with review questions. Draw the most stable enol from:

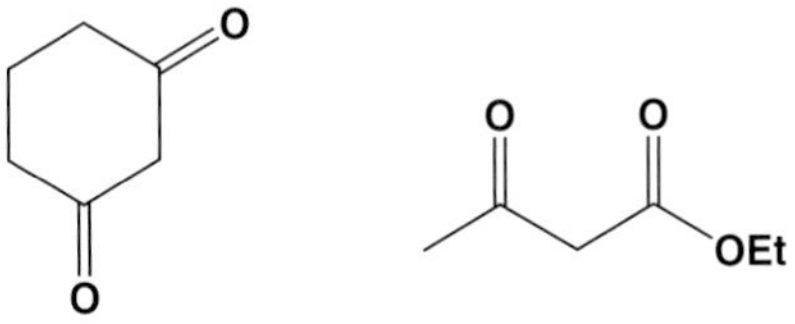

214.

215. Arrange these compounds in the order of reactivity towards:
        (i) nucleophilic addition,
        (ii) nucleophilic substitution, and
        (iii) enolisation:

216. The order is the same in each case:

        Most reactive: Acid chloride
                           Anhydride
                           Ester
        Least reactive: Amide

because each reaction involves negative charge being taken on the carbonyl oxygen atom:

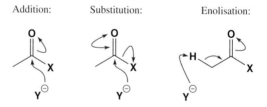

Addition: Substitution: Enolisation:

If you are in doubt about this, ask your adviser.

This is the end of the fourth section.

CHAPTER 5

BUILDING ORGANIC MOLECULES FROM CARBONYL COMPOUNDS

Contents

Chemistry of the Carbonyl Group:
A step-by-step approach to understanding Organic Reaction Mechanisms,
Revised Edition. Timothy K. Dickens and Stuart Warren.
© 2018 John Wiley & Sons Ltd. Published 2018 by John Wiley & Sons Ltd.

89

Concepts assumed:

- Relative reactivities of aldehydes, ketones, esters, acid chlorides and anhydrides.

Concepts introduced:

- Carbon–carbon bond formation from two components: enol and electrophile.

- Condensation reactions catalysed by base or acid.

- Selection of the right base for condensation reactions.

- Ambiguity of condensation reactions.

- Self-condensations and cross-condensations.

- Use of compounds with no enolisable protons.

- Methods of avoiding ambiguity in cross-condensations.

- Secondary amines as catalysts via $R_2C=N^+R_2$.

- Stability of five- and six-membered rings.

- Instability of four-membered rings.

- Use of β-dicarbonyl compounds to form stable enolates.

- Use of NO_2 and CN as carbonyl equivalents. Nucleophilic addition to enones.

- Decarboxylation and β-dicarbonyl cleavage reactions.

- Use of carbonyl groups as activating and directing groups in syntheses.

- Need to plan ahead in syntheses.

Concepts reinforced:

- Basic carbonyl mechanisms.

- Use of pK_a.

- Elimination of OH and R_2N in acid conditions.

- Similarity of reactions of acyclic and cyclic compounds.

217. In this part of the program, we are going to explore ways of making organic compounds from carbonyl compounds. The reactions of carbanions and enols in particular are some of the chief ways in which the organic chemist builds up complicated structures. Many people find this a daunting part of organic chemistry and so we have treated the subject in rather more detail than other subjects in the program. If you are reading this, you must have survived the earlier sections with some enjoyment and you ought easily to be capable of tackling this rather more demanding section.

218. Let's begin by stating very clearly just what possibilities we are considering. We have discussed *four* different types of reaction, which can occur between a nucleophile (base) and a carbonyl compound. List these in words.

219. They are as follows:

(a) Addition
(b) Substitution
(c) Complete removal of carbonyl oxygen
(d) Removal of an α-proton: enolisation.

One possibility we have not yet considered is combining (d) with the other three: that is, using the enolate anion as the nucleophile in these reactions.

Make the enolate anion from acetaldehyde and add it to the carbonyl group of another molecule of acetaldehyde.

220.

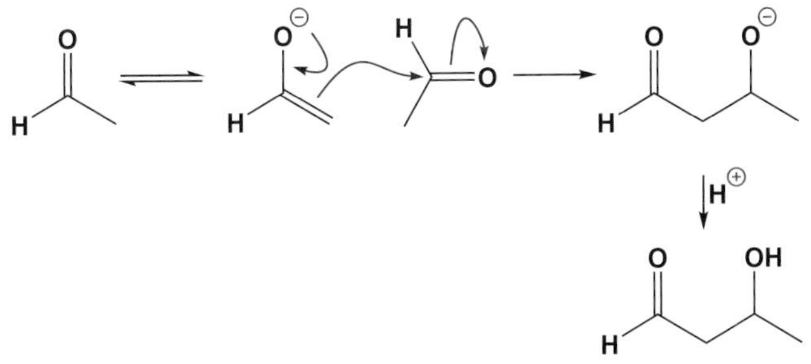

In making a four-carbon chain from two-carbon units, we used a nucleophilic enolate and an electrophilic carbonyl group. This theme of the two components is universal to this section.

---

221.   Now make the enolate from ethyl acetate and attack another molecule of ethyl acetate with it.

---

222.   Have you just added the one to the other, or have you extended the reaction, as you should, to make it a substitution?

---

223.   Here, then, is the full reaction:

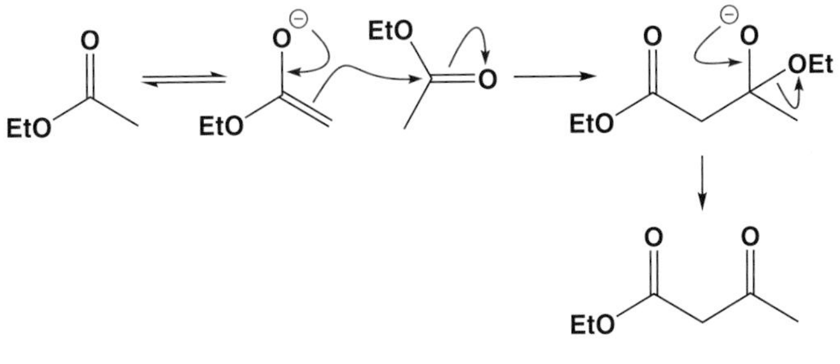

---

224. These reactions were catalysed by an unspecified base. With the aldehyde, quite weak base is needed and there is no possibility of an alternative reaction, but we couldn't use hydroxide ion, for example, to enolise the ester. Why not?

---

225. Hydroxide ion would attack the carbonyl group and hydrolyse the ester. We have an ingenious way of getting round this problem. There is one base, and one only, that we can use to enolise ethyl acetate so that, if it *does* attack the carbonyl group and displace EtO⁻, we don't care. What base is that?

---

226. Clearly EtO⁻!

If it attacks the carbonyl group, it simply makes another molecule of ethyl acetate and we don't notice:

---

227. These reactions can also be catalysed by acid. Try your hand at enolising acetone in acid solution and then using the enol to attack another molecule of acetone, still in acid.

228.

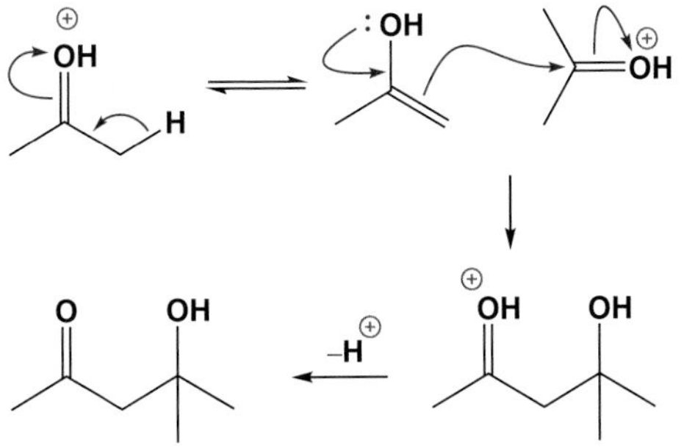

You may have carried this reaction a step further. If you didn't, consider now what would happen to that product under the reaction conditions.

---

229. The hydroxyl group will be protonated and lost:

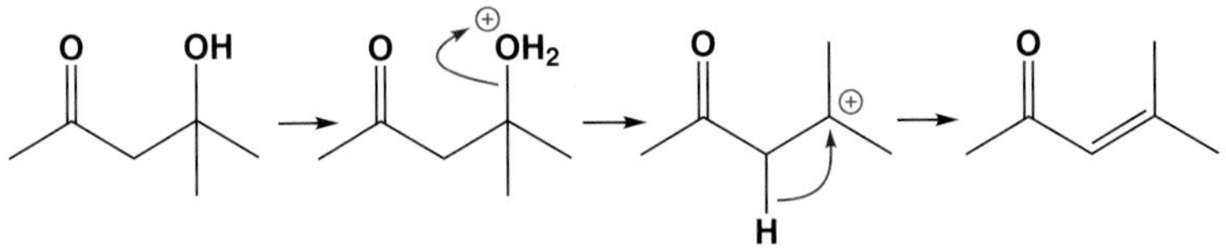

Have you noticed that we have now fulfilled the specification of *frame* **219**? We have used enols as nucleophiles in reactions of types (a), (b), and (c).

---

230. Reactions of this type in which carbonyl compounds combine together to form longer carbon chains are (loosely) known as condensation reactions. So far, we have considered only *self-condensations*, that is, ones in which both the nucleophilic component and the electrophilic component are forms of the same molecule. How would you make this compound?

---

231. By self-condensation of an ester, using EtO⁻ as base:

What about this?

232. By self-condensation of cyclohexanone, using acid or base as catalyst:

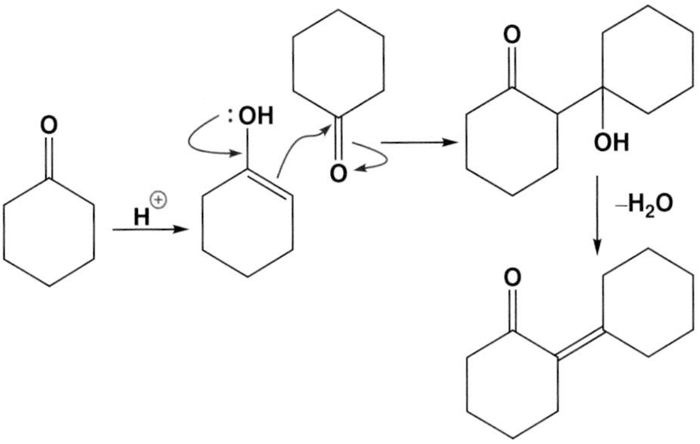

---

233. Small structural changes can easily be introduced after the basic skeleton is complete, a process often known as *elaboration*. How would you synthesise this compound?

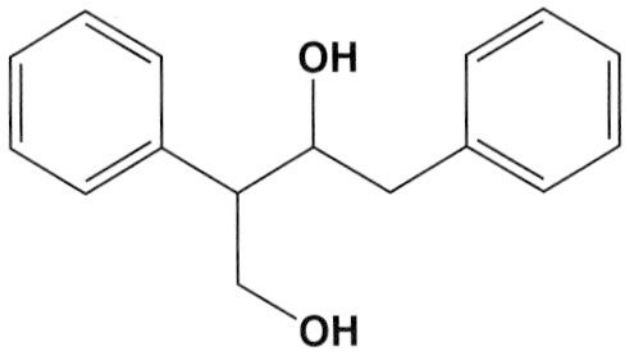

---

234. The basic skeleton is the same as the one we made in *frame* **230**. All we have to do is to reduce both ester and ketone, and the reagent for this is LiAlH$_4$ (*frames* **43–45** and **78–79**).

---

235. So far, we have used the same molecule to provide both nucleophilic (enolate) and electrophilic (carbonyl) components. This avoids ambiguity, but we must also be able to do *cross-condensations* in which different molecules supply the different components. What problem would arise if we tried to do this condensation?

236. Though we have labelled them, the molecules can't be expected to know which role they're supposed to fulfil and may give products of self-condensation.

This is a good place to rest if you want to.

237. One way round this problem is to use components that can't *self-condense*. Select from this list some molecules that fit the bill:

(a)	*t*-BuCHO	(b)	Me(CO)OBu^t	(c)	PhCHO
(d)	EtO(CO)OEt	(e)	CH$_2$=O	(f)	H(CO)OEt
(g)	PhCH$_2$CHO	(h)	EtO(CO)(CO)OEt		

---

238. All except (b) and (g) cannot self-condense because they have *no enolisable protons*. These are very useful compounds. Make a classification of the type of substituent we can attach to a carbonyl group so that it can't self-condense. Our list has five types.

---

239. Here is our list; yours may be different:

(1) tertiary alkyl groups like Bu^t.
(2) aryl groups like Ph.
(3) hydrogen, H.
(4) electronegative substituents like Cl, OEt.
(5) other carbonyl groups as in *frame* **237h**.

With a list like this, we can invent molecules that can't enolise. Write down a ketone and an acid chloride that can't enolise.

---

240. There are lots of answers such as:

Ketone:

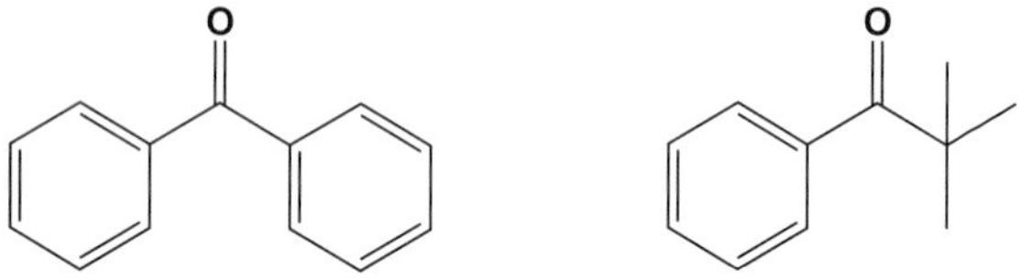

Acid chloride:

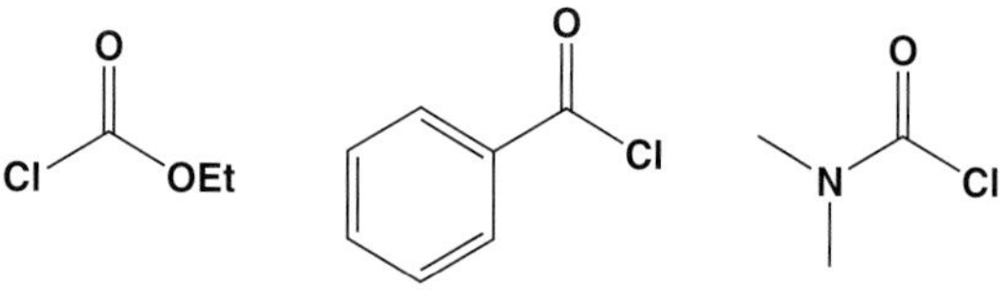

One of these compounds can act only as the electrophilic component in a condensation and can't self-condense; what will be formed here?

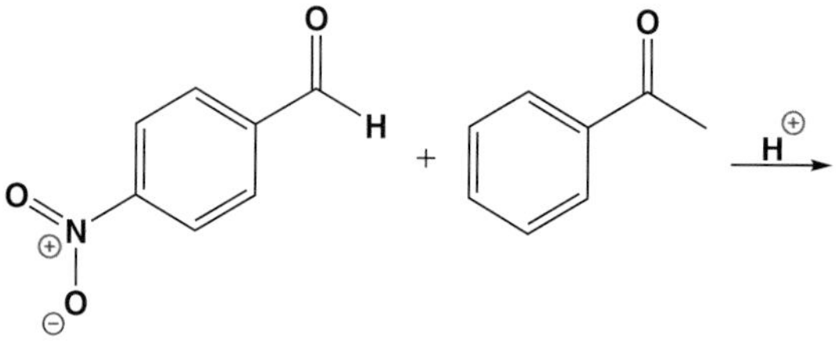

241. The only compound that can enolise is the methyl ketone:

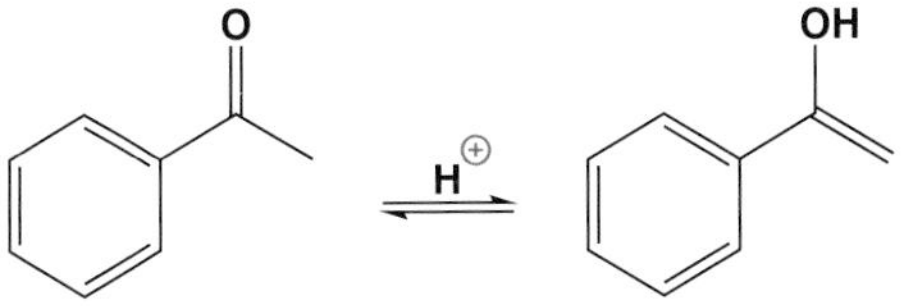

and it will then attack the aldehyde, which can only act as the electrophilic component:

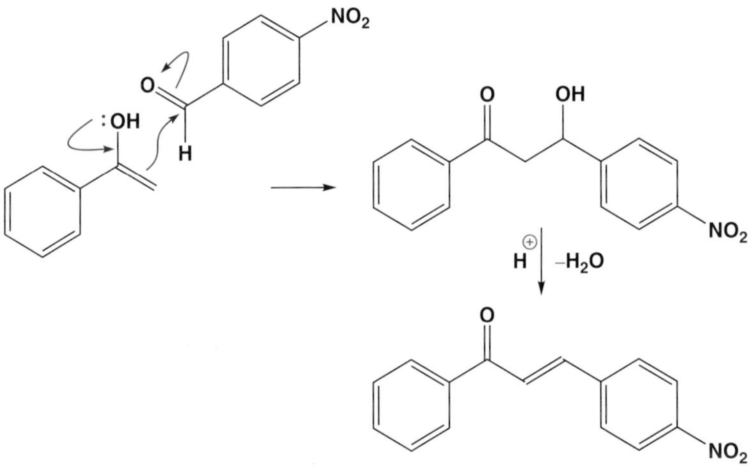

242. There was in fact one ambiguity left in this last example. The aldehyde could act only as the electrophilic component, but the ketone could act as both components, that is, it could self-condense. In fact, it didn't. Why not?

---

243.   If you don't see this at once, imagine what happens when a molecule of ketone enolises. It then begins to look around for an electrophilic molecule, and it has two choices: another molecule of itself or the aldehyde.

---

244.   The point here is that the aldehyde is more reactive towards nucleophiles than the ketone: so, when the ketone enolises, it will always attack the more reactive aldehyde, and never the less reactive ketone.

Aldehydes, compounds with two adjacent carbonyl groups, and formate esters (H(CO)OR) are suitable reactive compounds for this type of condensation. What would be formed here?

---

245.   The ketone alone can enolise, and the enol will prefer to attack the formate ester:

---

246.   Here, therefore, we have our first safe method of doing a cross-condensation. We use any enolisable compound as the nucleophilic component, but what we must have as the electrophilic component?

---

247. A compound that can't enolise, but which is more reactive than the other compound towards nucleophiles.

---

248. One useful compound for this type of reaction is formaldehyde, $CH_2=O$, and we often carry out condensations with this using a secondary amine and dilute acid. The amine gives this reaction:

$$CH_2O + R_2NH \xrightarrow{H^+} CH_2=N^+R_2$$

Draw a mechanism for this.

249.

This reaction is very like the condensation of primary amines with ketones (*frames* **123–125**). If we have Ph(CO)Me present in the solution, what will happen now?

---

250. $CH_2=N^+R_2$ is of course very electrophilic and picks up the enol from the ketone:

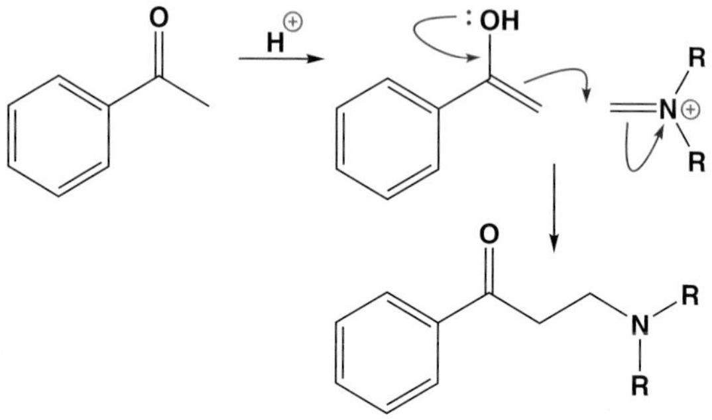

This is known as the Mannich reaction. What product would be formed here?

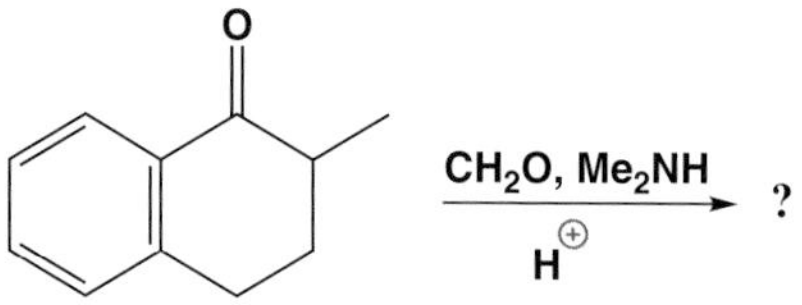

251.

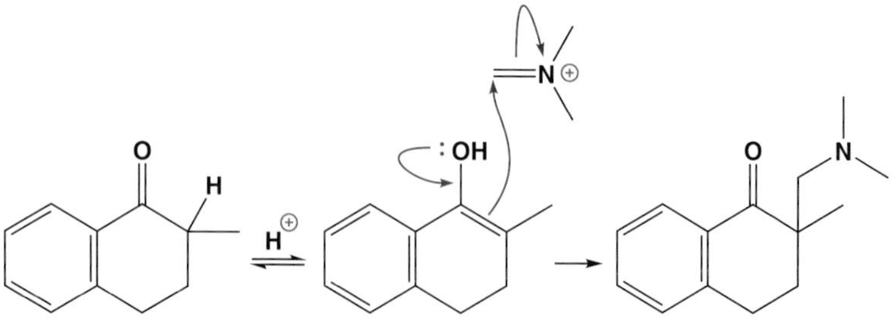

252. A kind of enolic component we haven't mentioned yet is the acid anhydride. If you wanted to make the enolate anion from acetic anhydride, what base would you recommend?

---

253. We would use acetate ion as base, since any substitution reaction then simply regenerates the anhydride. Using the anhydride then as the enolic component, what happens with PhCHO as the electrophilic component?

254.

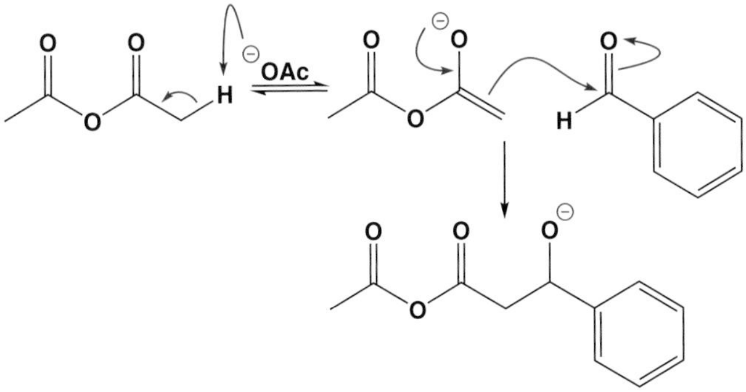

The O⁻ is ideally placed to attack one of the carbonyl groups in the molecule intramolecularly. Can you draw this?

---

255. It could attack the nearer carbonyl group making a four-membered ring:

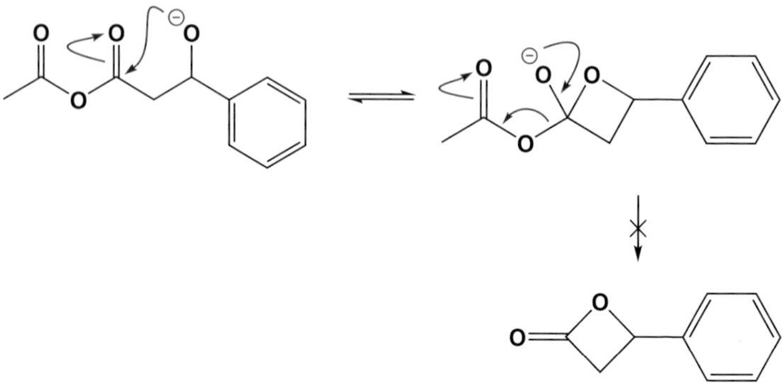

or the other making a six-membered ring. As you know (*frames* **95ff.**), six-membered rings are very stable, and this is the preferred reaction:

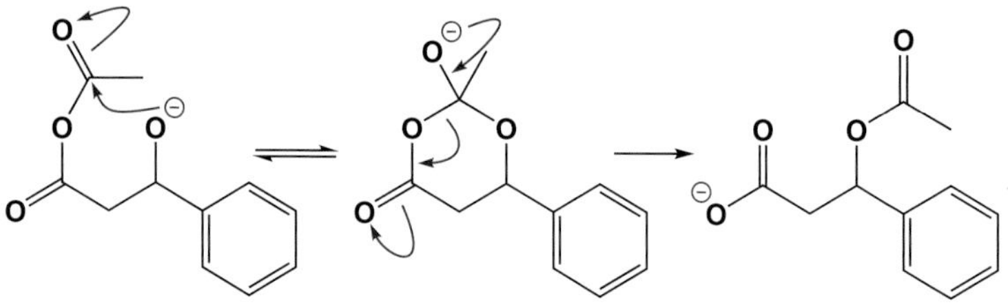

256. Under the conditions of the reaction, an elimination occurs:

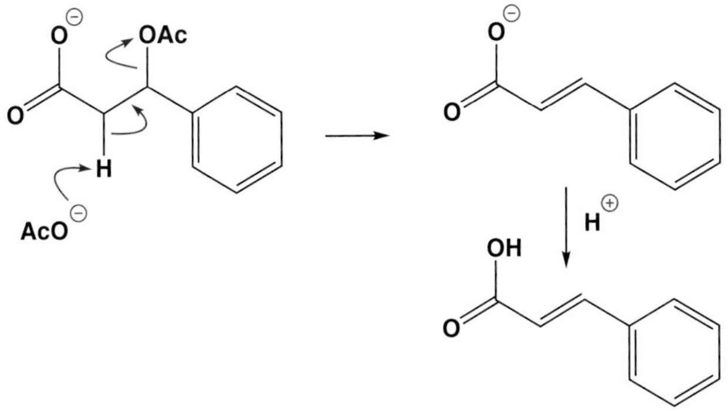

and acid workup gives as a final product an α–β unsaturated acid. How would you make this compound?

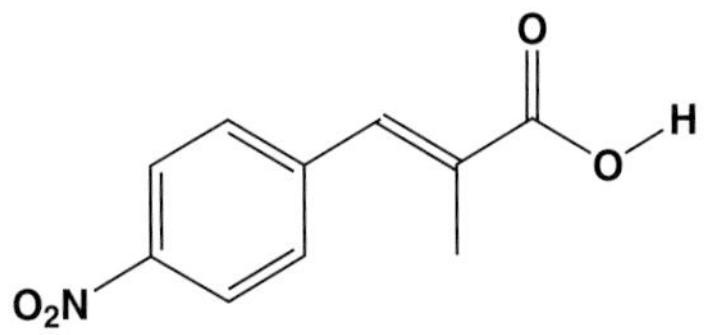

_____

257.

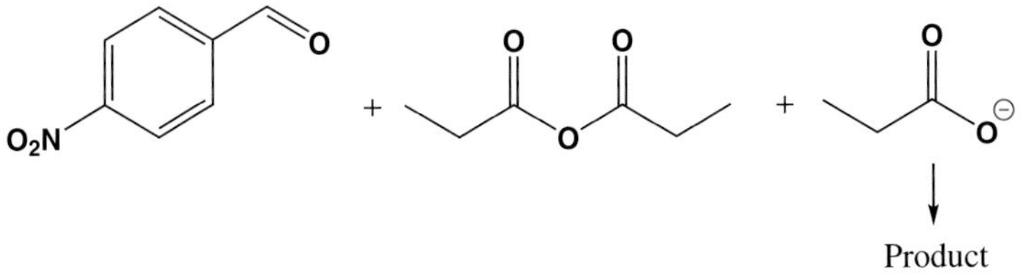

Product

Note that with the change of acid anhydride, a change of base is called for.

_____

258. Just to check that you've understood these ideas, what would happen if we tried to condense acetaldehyde as enolic component with benzophenone ($Ph_2C=O$)?

---

259. Acetaldehyde would enolise alright, but it would never react with $Ph_2C=O$. It would much prefer to react with another molecule of itself:

This is a good place to rest if you want to.

---

260. We have done a number of cross-condensations so far by using a reactive electrophilic component, which doesn't act as a nucleophile because it can't enolise. Now we're going to look at the reverse: a reactive enolic component that doesn't act as an electrophile. We thus want to increase the ability of a compound to enolise without making the carbonyl group more electrophilic. One way to do this is to use two carbonyl groups, both of which encourage the enolisation of the same CH group. Using ester functions, draw such a compound.

---

261. The simplest compound of this sort is

What base would we use to make the enolate anion here?

---

262.   Ethoxide ion, so that it doesn't matter if substitution occurs. We shall get a substantial amount of enolate here because the pK_a of EtOH is about 16 and that of the CH in *frame* **261a**.

Select which of the following compounds would form substantial amounts of enolate anion with ethoxide ion in ethanol. List of compounds:

(a)   Ph(CO)Me   (b) CH_3NO_2
(c) $CH_3(CO)CH_2(CO)OEt$ (d) $H_2C=O$
(e) Bu^fCHO (f) $CH_2((CO)OEt)_2$
(g) $EtO(CO)CH_2CN$ (h) $((CO)OEt)_2$
(i) $PhCH_2CHO$ (j) $Ph(CO)CH_2(CO)CH_3$

---

263.   (b),(c),(f),(g) and (j) are reactive enough. If you got all these and included no others, skip to *frame* **268**. Otherwise read on: if you included (a) or (i) read *frame* **264**; if you included (d), (e), or (h), read *frame* **265**; if you omitted (b), read *frame* **266**; if you omitted (c), (f), (g), or (j), read *frame* **267**.

---

264.   To get a substantial amount of enolate anion, the pK_a of the carbonyl compound wants to be several pH units below that of the alkoxide ion. In fact the limit is about 14 using the ethoxide ion. The compounds you included are therefore too weakly acidic with pK_a values of about 20 (a) and 18 (i). Go to *frame* **268**.

---

265.   Yes, these compounds are very reactive in the electrophilic sense, but they have no enolisable hydrogen atoms. Go to *frame* **268**.

---

266.   It is rather surprising to find a compound with only one acti-vating group among the others, but the nitro group is very electron withdrawing so that the pK_a of nitro-methane is 10: it is even more acidic than most dicarbonyl compounds. Go to *frame* **268**.

---

267.   The chief group of compounds acidic enough to provide sub-stantial amounts of enolate ion with ethoxide base is the dicarbonyl compounds. They have pK_a values in the regions 9–13.

---

268.   Draw the enolate anion from cyanoacetic ester EtO(CO)CH$_2$CN and show why it is stable.

269.

Now use this anion to attack a molecule of acetone.

270.

In practice, this reaction is usually done with catalysis by a combination of a secondary amine and acid. Can you suggest a role for the secondary amine?

---

271. It combines with the electrophilic component (acetone in this case) to make it even more electrophilic. We met this previously in the Mannich reaction (*frames* **248–250**).

Under these acidic conditions, a further reaction will take place. Can you suggest what?

---

272. If you don't know, look at the product and see where protonation is most likely to occur: one of the groups is protonated and then falls off.

---

273.   An elimination reaction occurs:

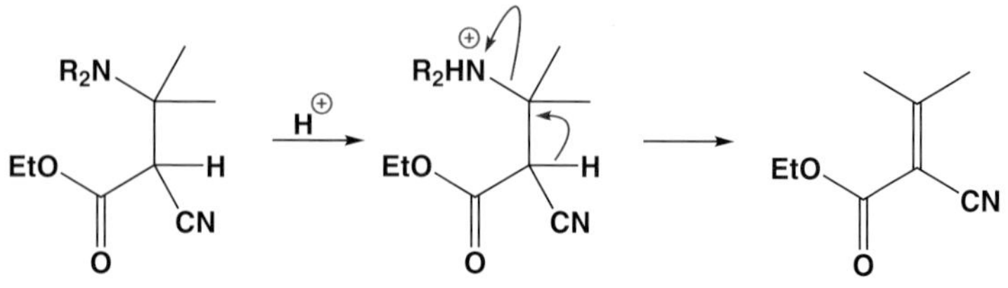

This reaction goes under the name of the Knoevenagel reaction. How would you make this compound?

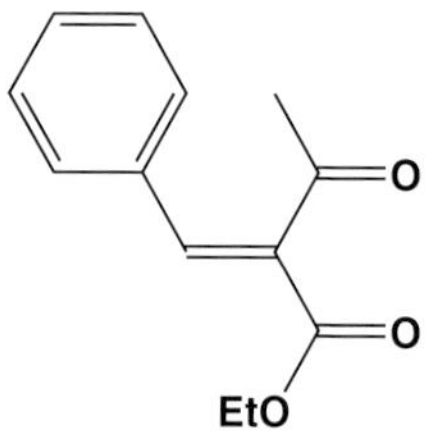

---

274.   By a Knoevenagel reaction between benzaldehyde and ethyl acetoacetate:

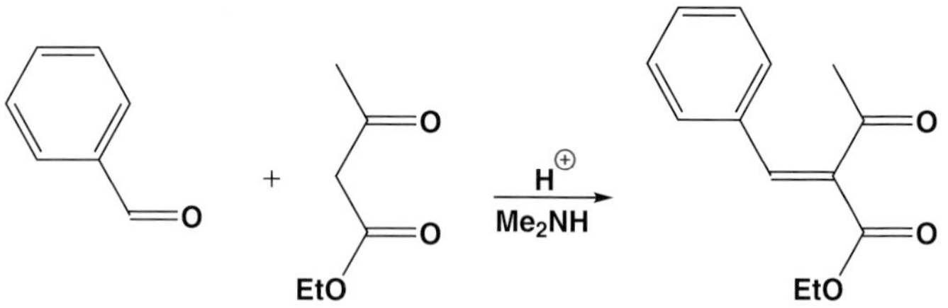

---

275. The enolates from β-dicarbonyl compounds are so easily formed that they can be used in a very simple carbon–carbon bond-forming reaction outside our general scheme. Consider what would happen if you made the enolate anion from the compound below and reacted it with methyl iodide:

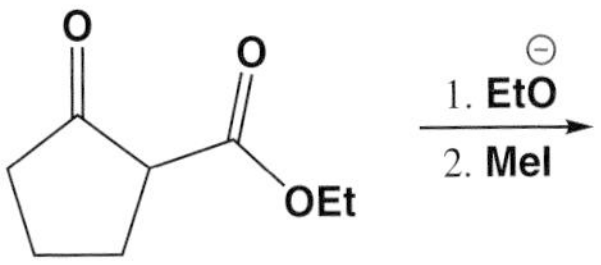

276. The enolate anion attacks methyl iodide in an S_N2 reaction; in other words, it is alkylated:

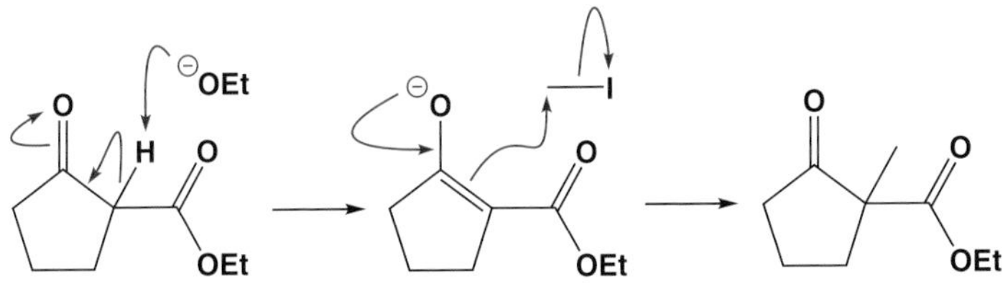

This reaction extends the scope of our work with β-dicarbonyl compounds considerably. Do you remember how we synthesised them in the program? How, for example, would you make this?

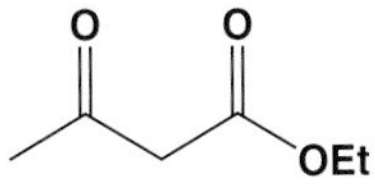

277. By a self-condensation reaction of an ester:

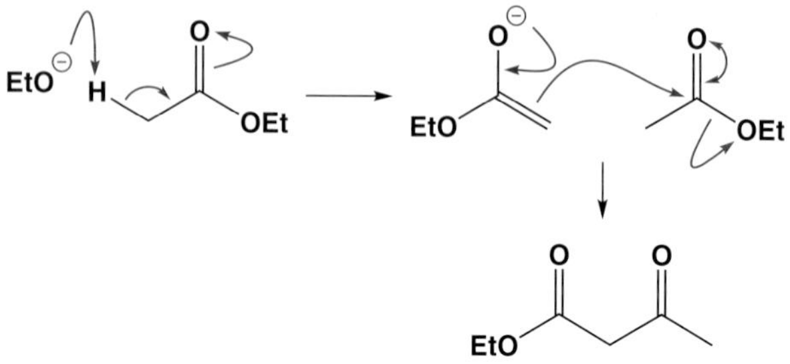

How, then, would you make this?

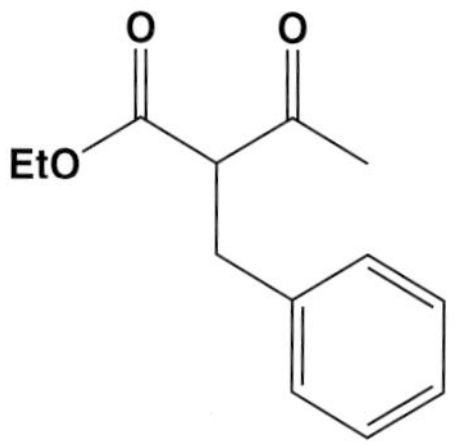

_____

278. First make the right β-dicarbonyl compound and then alkylate it with the right alkyl halide:

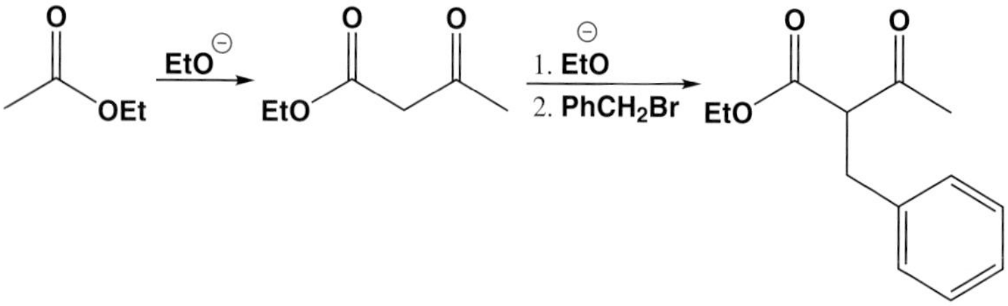

_____

279. We might summarise our syntheses so far in terms of the general classes of molecules we are able to make. Look back over the program (*section* **5**) and see if you can make a short list.

---

280. We can make:

i. β-hydroxycarbonyl compounds by addition

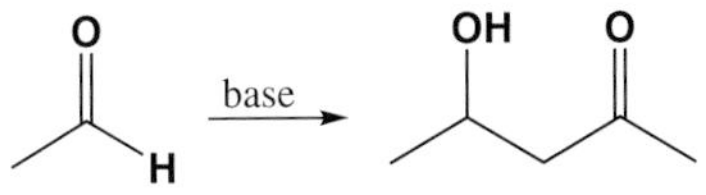

ii. α–β unsaturated carbonyl compounds by loss of water from the β-hydroxy compounds:

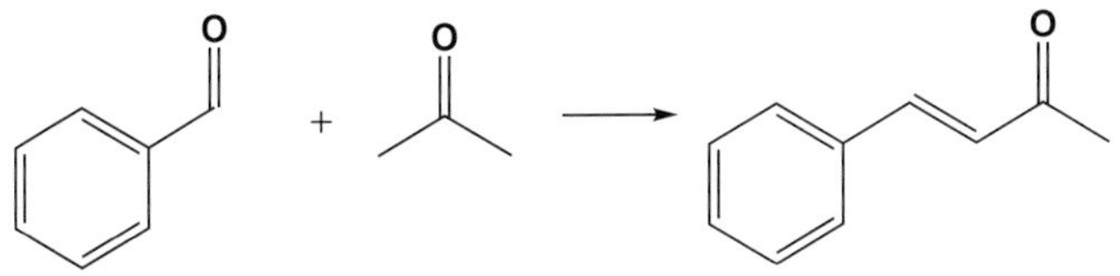

iii. β-dicarbonyl compounds by substitution:

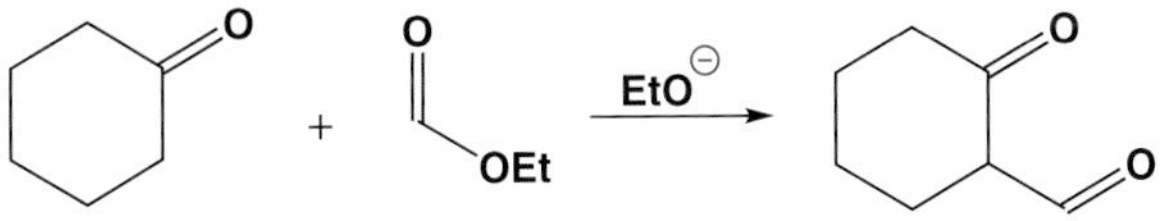

Have you noticed that in all these compounds there is the same relationship between the two original carbonyl groups? The one originally supporting the enol and the electrophilic one always end up in a 1,3-relationship. Mark these two carbon atoms (★) on each of the products in this frame.

281.

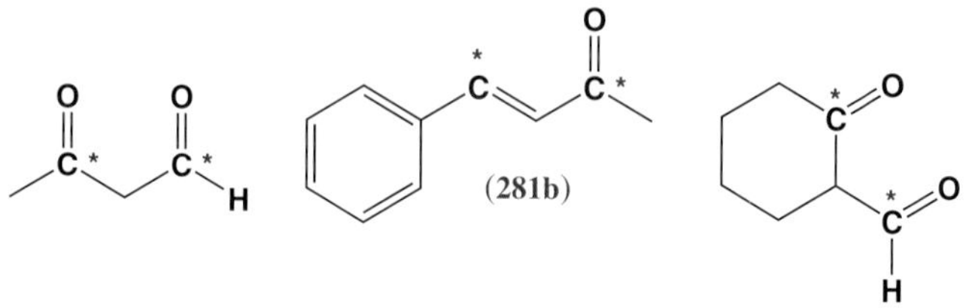

(281b)

So, at the moment, the scope of our skill at synthesis is rather limited. We can extend it by the use of α,β unsaturated ketones as electrophiles. Where would the compound we have just made **(281b)** be attacked by nucleophiles?

---

282. Have you shown two positions?

283.

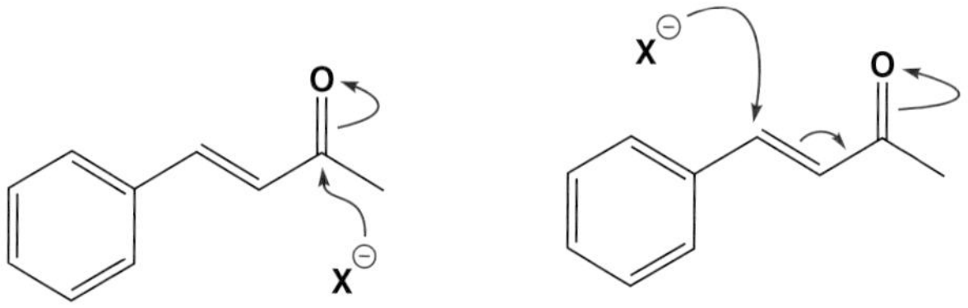

Fortunately, enolates usually prefer to add to the end of the C=C double bond to give the more stable anion. Add the enolate from $CH_2((CO)OEt)_2$ to this ketone.

284.

This intermediate is of course another, less stable, enolate anion, and in the ethanolic solution it will protonate to give the final product. Draw this.

285.

Notice that the compound we have just made has a 1,5-relationship (★) between the two carbonyl groups. This is an example of the Michael reaction. Here is another: suggest what the product would be.

286.

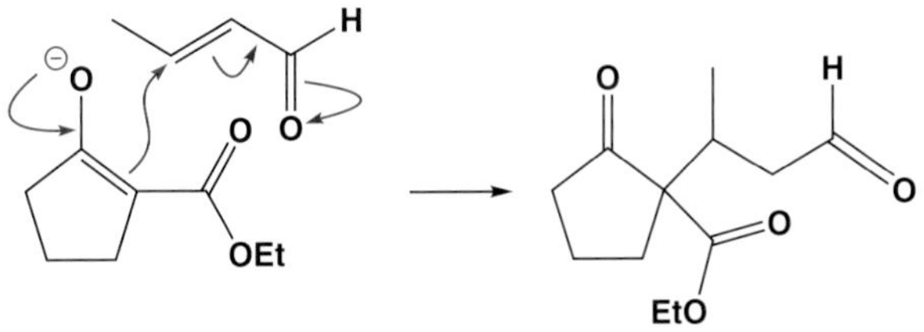

In order to use the Michael reaction, we need to be able to make α,β unsaturated carbonyl compounds of various types, but we already know how to do this. How would you make this one?

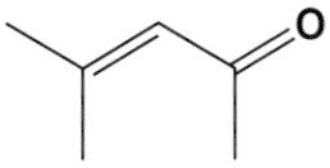

---

287. By the self-condensation of acetone:

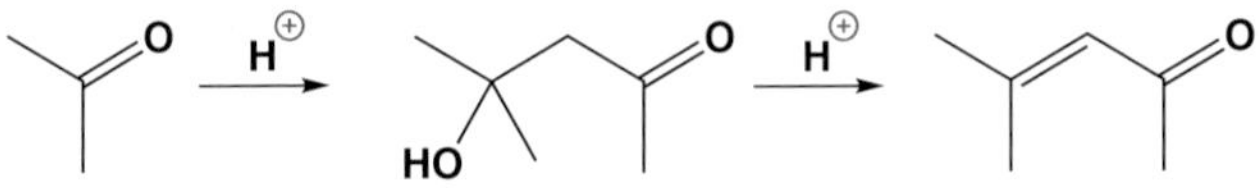

In fact, the self-condensation of aldehydes and ketones is a good way to make Michael reagents. We can't, however, make α,β unsaturated esters this way. How can we make this, then?

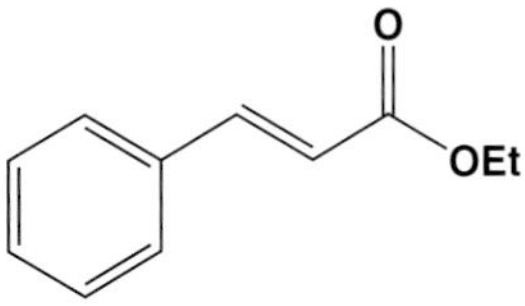

---

288.   There are three methods that we have met so far:

i. Ester and carbonyl compounds that can't enolise:

ii. Perkin reaction (*frames* **252–257**)

iii. Organo-zinc reagents (*frames* **208–212**)

289. A compound we have mentioned but not used is nitromethane (*frame* **262**), a good source of a stable enolate. What product would we get here?

290.

291. Another non-carbonyl activating group is cyanide, which withdraws electrons in much the same way as carbonyl. Acrylonitrile, $H_2C=CHCN$, is a useful Michael reagent. What would happen here?

292. Michael addition of enolate anion to acrylonitrile:

293. To finish off the Michael reaction, let's take two compounds we've already made (*frames* **230–232**) and combine them together:

294.   Yes, we know this is a very complicated example, but use your knowledge: remove a proton from the most acidic position, and add the resulting anion, Michael fashion, to the other compound.

This example is fictional, that is, to the best of our belief, no one has tried it, and maybe you don't blame them. We put it in just to show you that the complicated molecules can be made this way. Just suppose someone asked you how to make this molecule. We imagine you would find it a rather daunting problem. Yet these are reactions and compounds we have ourselves made in this program with straightforward ideas.

This is a good place to rest if you want to.

295. It must be clear to you by now that the carbonyl group not only allows certain reactions to occur which would otherwise be impossible: it also directs where they will occur by selecting exactly which hydrogen atom shall be removed in enolisation and therefore exactly where the new carbon–carbon bond is to be formed.

Organic chemists often introduce *unnecessary* carbonyl groups into molecules during a synthesis, just so that they can use this direction-finding function. These groups are not wanted in the final molecule and so we must be able to remove them later. We have just been discussing this molecule:

What would happen if we boiled this up in water containing a catalytic amount of acid?

296. The ester group would hydrolyse:

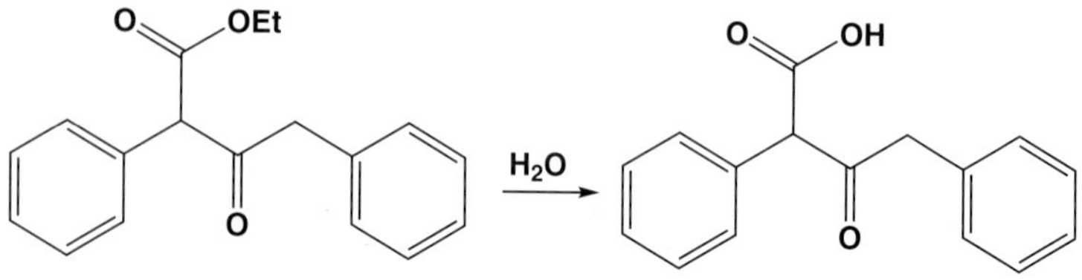

and as soon as the free carboxyl group was released it would be released in another sense in that it would form CO_2. Arrows?

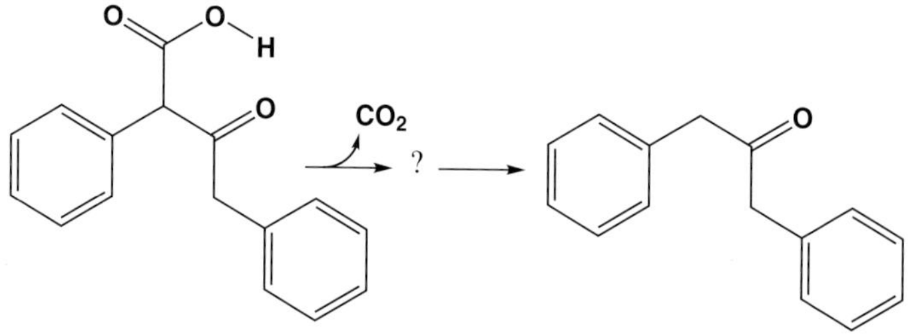

297. The missing intermediate is the enol:

The decarboxylation reaction happens to this sort of compound too:

From *frame* **285**:

Mechanism?

298.

Hydrolysis of both ester groups is followed by decarboxylation of one of them to give the enol of the final product. Decarboxylation is a useful addition to the alkylation reactions we mentioned in *frames* 275–278. What happens here?

299.   First we alkylate the enolate anion:

then hydrolyse and decarboxylate the ester:

300.   If we combine decarboxylation with the Michael reaction (*frames* 281–285), we get a general synthesis of 1,5-diketones. Complete this reaction?

301.   The β-dicarbonyl compound will enolise, as usual, and add to the Michael reagent:

Now we must hydrolyse the ester and the usual decarboxylation follows:

302. A more deep-seated change than decarboxylation can easily occur. You will remember that we have stressed all along that these carbonyl reactions are reversible, though we don't always show them as such. Even the formation of carbon–carbon bonds is reversible. If we take some of this compound and treat it with EtO⁻, will it form a substantial amount of enolate?

---

303. No, because, although it can form an ordinary enol,

the two carbonyl groups can't cooperate in forming a very stable enol as the position between them is blocked. What happens instead is that EtO⁻ attacks the more reactive carbonyl group to give an adduct. Draw this.

---

304. The ketone carbonyl is more reactive than the ester carbonyl:

The adduct then decomposes with the loss of the enolate anion of the ester as a leaving group. Draw this.

305.

If the compound had been able to form a substantial amount of enol, it would have done this much less readily. We can generalise from this reaction and say that under the right conditions we ought to be able to *undo* any of the C—C bond-forming reactions we have been using. What would happen here?

306.   This time the aldehyde is the more reactive group:

These examples show how we can build-up a molecule in disguise and reveal it at the last minute by a specific cleavage reaction. The point of leaving the cleavage reaction until the end is that we retain the 1,3-relationship of the two carbonyl groups and can thus control the way the molecule reacts. These cleavage reactions are even more dramatic when they are applied to cyclic compounds, and it is time for us to look at the formation and cleavage of rings in more detail.

---

307.   We have met six-membered rings several times during the program and noted their stability. This is a property they share with five-membered rings.

We have also met this β-keto ester, but we haven't yet worked out how we would synthesise it. Suggestions?

308. We made β-keto esters in the past by the self-condensation of esters (*frames* **221–230**). This molecule is also made by the self-condensation of an ester.

---

309. The reaction is intra-molecular: the enolate component being one end and the electrophilic component the other end of a di-ester of a six-carbon acid:

310. In cyclic compounds, the problem of ambiguous reactions often just doesn't arise because the molecule will always prefer reactions that form five- or six-membered rings to those which form other sizes. Four-membered rings are particularly bad. Take this diketone, for example:

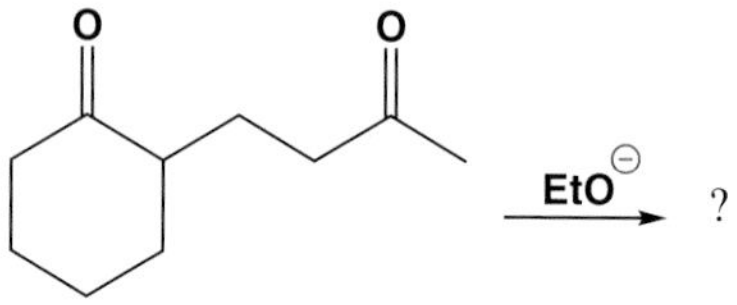

It can form four different enolates, each of which could react with the other carbonyl group, and yet only one product is formed. Which?

---

311. Two possible reactions give four-membered rings, and can be discounted, and one reaction gives a good, stable, six-membered ring. There is actually another possibility that you could discuss with your adviser if you are interested. It is beyond our scope now.

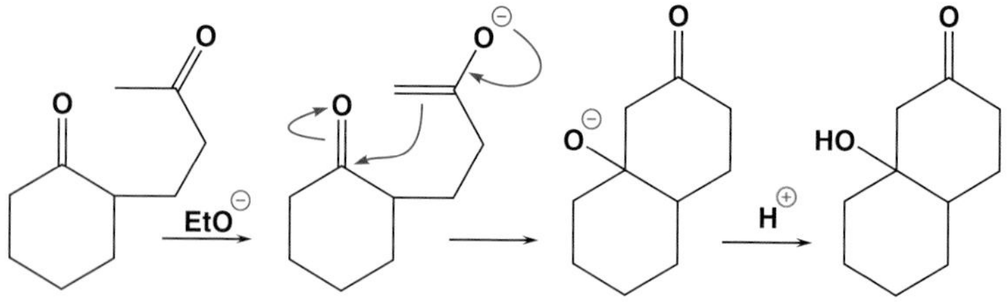

312. Quite often these cyclisation reactions happen spontaneously after some other normal reaction. Carry out the normal Michael reaction between these two compounds and then see if you can predict what cyclisation will follow:

313.   Here is the Michael reaction:

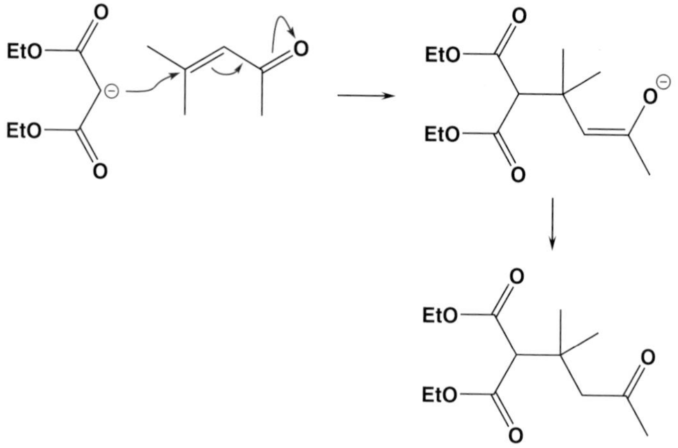

and here is the cyclisation: the only one to give a six-membered ring. How could we remove the COOEt group from this product, and what would we get?

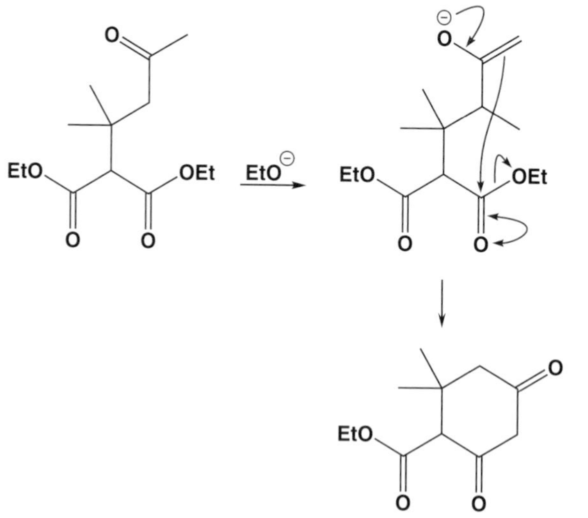

314. By acid hydrolysis: decarboxylation would follow:

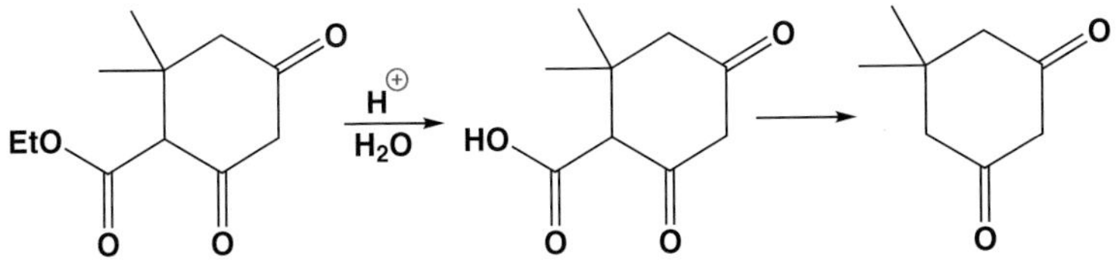

Try your hand at another reaction like this: a reaction between two compounds that is followed by cyclisation; this time a five-membered ring is formed:

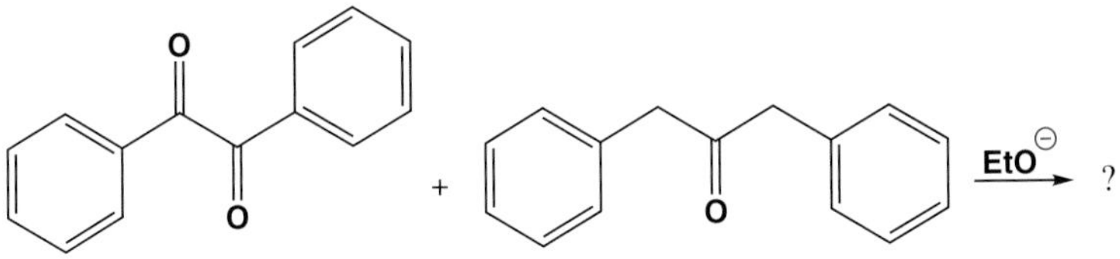

315. Only one compound can enolise, and the diketone will act as the electrophilic component as it is more reactive than a mono-ketone. We begin with a simple condensation:

Now we are all set up for an intramolecular condensation:

---

316. The cleavage reactions we talked about a few frames ago (302–306) can also occur in cyclic compounds and we may get ring opening here. What might this lead to?

317.   The ketone carbonyl is the more reactive so EtO⁻ adds there. The only leaving group is the ester enolate:

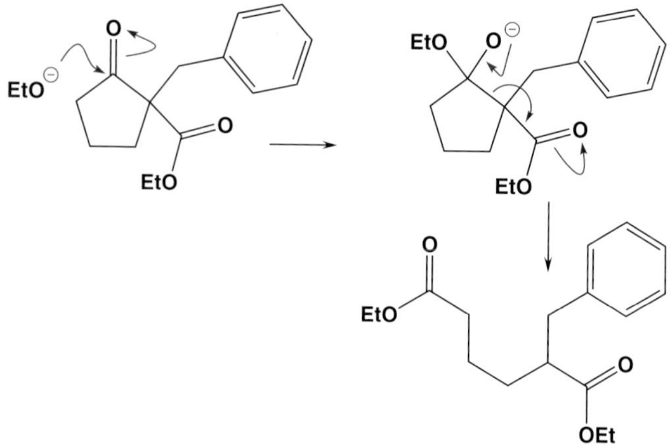

318.   Sometimes ring opening can be caused by other reactions than base cleavage. Predict what happens here in a reaction that contains a bit of revision work:

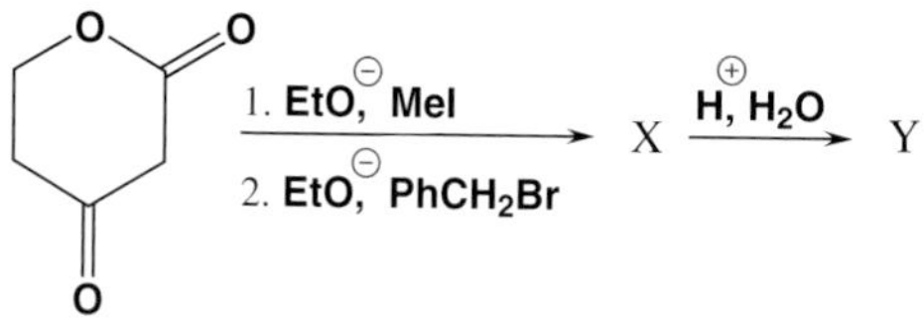

319. The first two steps are simply alkylations (*frames* **275–278**):

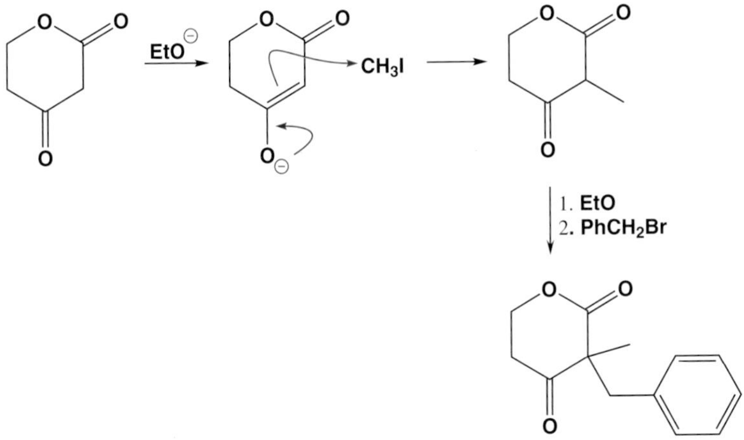

Then acid hydrolysis opens up the cyclic ester and decarboxylation follows:

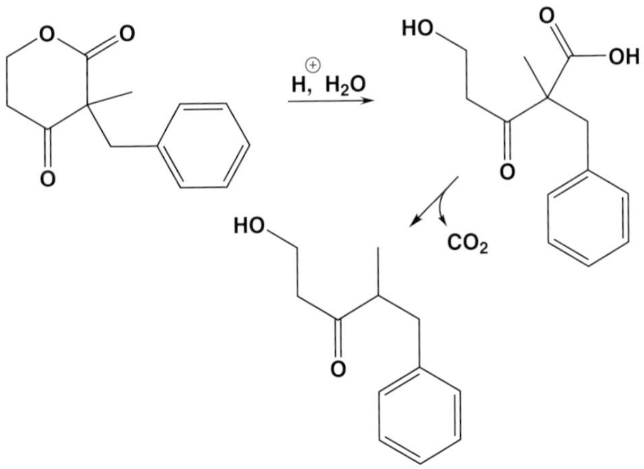

The function of the ester carbonyl group was simply to introduce the two alkyl groups: after it had done this it could easily be removed.

This is a good place to stop and rest if you want to.

320. We have now completed the exploration of new territory, and we should stop here a moment to consolidate the new ideas. The main ideas have been ways to make new carbon–carbon bonds between two compounds, both having a carbonyl group, but each acting in a different way. What do we call these two reagents?

---

321. One is the nucleophilic or enolic component. The other is the electrophilic or carbonyl component. We have discovered various ways to join these components unambiguously. Make a list of these.

---

322. Have you remembered to include reactions in which both components are the same molecule and reactions that form rings?

---

323.   One way to express this is:

(a) Self-condensation: both components derived from the same molecule (*frame* **230**).

(b) Electrophilic component incapable of enolising and more reactive than the enolic component towards nucleophilic attack (*frames* **240–244**).

(c) Enolisable component is very reactive in the sense that substantial amounts of enol are formed (mostly β-dicarbonyl compounds) but unreactive in the electrophilic sense (*frames* **260ff**).

(d) A six- or five-membered ring is formed. Other possible reactions led to less stable rings of other sizes (often four-membered) (*frames* **310–311**).

You may well have expressed this just as correctly but in a different way. If you are in doubt, discuss with your adviser.

---

324. The rest of the program will try to put these ideas into perspective by looking at some actual syntheses, offering some general problems, and generally giving you a hint of the enormous possibilities in these simple ideas. Don't be put off if you find the going a bit rough: you've really done all the hard work by this stage!

Here is a step from the synthesis of a steroid, one of a large family of physiologically active molecules, including hormones. Explain what is happening.

325.   It is clear that the CHO group has been lost, presumably in a β-dicarbonyl cleavage reaction (*frames* **305–306**), and that a condensation has taken place to form a new six-membered ring. The order of these two reactions doesn't matter much:

326.   Here is a part of another steroid synthesis. This time, see if you can suggest how to carry out the steps marked ★.

327.

A ⟶ B:	Cross-condensation of type **323b**.
B ⟶ C:	Simple alkylation of stable enolate anion.
D ⟶ E:	Hydrogenation of a double bond.
E ⟶ F:	Steps outside the scope of this program.
F ⟶ G:	Cyclisation to form stable five-membered ring.
G ⟶ H:	Hydrolysis and decarboxylation.

You may be interested to see the full steroid skeleton, and here it is. Perhaps you can see that these two syntheses are moving towards it, though both still need many steps.

A steroid:

328.   We shall end the program with one complete synthesis: that of cedrene, the essential oil of cedar wood. The synthesis has many steps and requires some reactions you haven't met, but we thought you would like to see the carbonyl group in action. Here it is:

You should be able to suggest reagents for the steps marked ★ and mechanisms for the steps marked **M**. Don't worry about the other stages in the synthesis as they use reagents you haven't met.

329. Here are outline answers: if you're in doubt about any of them, discuss them with your adviser.

A ⟶ B:	Alkylation of a stable enolate.
B ⟶ C:	Decarboxylation after nitrile hydrolysis.
C ⟶ D:	Cyclisation to form a five-membered ring: all the other possible reactions give three- or four-membered rings.
D ⟶ E:	Make the stable enolate with EtO⁻ and alkylate with MeCHBr(CO)OCH₂Ph.
F ⟶ G:	The methyl group on the ketone carbonyl is the enolate component attacking the ring ketone and forming a stable five-membered ring.
H ⟶ I:	Make the cyclic ketal with HSCH₂CH₂SH and hydrogenate over Raney nickel (*frame* **146**).
J ⟶ K:	A fine reaction, this! The methyl group on the ketone makes an enolate, and this attacks the ester group in a substitution reaction. Though the new ring looks a bit awkward, it is in fact a stable six-membered ring.

330. We stopped this synthesis when we had made the cedrene skeleton, and there are several stages still left before cedrene itself is formed. Cedrene is in fact:

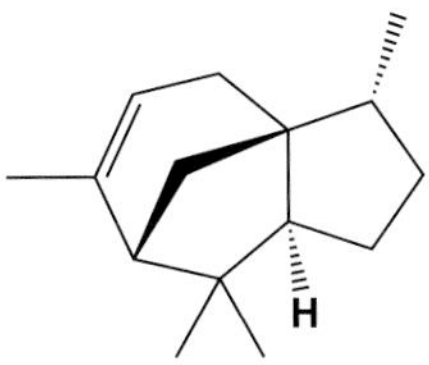

so that all the carbonyl groups used during the synthesis, a small matter of four ester functions and three ketone functions, were put in to guide the synthesis — the final product contains no carbonyl groups at all! This synthesis was designed by Professor Gilbert Stork and the genius in the plan is clear: every time he wants to make a carbon–carbon bond he has carefully arranged for the presence of two carbonyl groups, one to stabilise the enol and one to be the electrophile. These are the principles we have been studying here.

This is the end of the program.

Index

Chemistry of the Carbonyl Group: **155**
A step-by-step approach to understanding Organic Reaction Mechanisms,
Revised Edition. Timothy K. Dickens and Stuart Warren.
© 2018 John Wiley & Sons Ltd. Published 2018 by John Wiley & Sons Ltd.